GROUND WATER MONITORING TECHNOLOGY

PROCEDURES, EQUIPMENT AND APPLICATIONS.

GROUND WATER MONITORING TECHNOLOGY

PROCEDURES, EQUIPMENT AND APPLICATIONS.

BY ROBERT D. MORRISON

PUBLISHED BY
TIMCO MFG., INC.
PRAIRIE DU SAC, WISCONSIN

Published by Timco Mfg., Inc.
851 Fifteenth Street, Prairie du Sac, Wisconsin 53578

Designed and produced by
Tony's Studio, Sauk City, Wisconsin 53583

Printed and bound in the United States of America
First Printing 1983

Morrison, Robert D.
Ground Water Monitoring Technology.
Procedures, Equipment and Applications.

Library of Congress Catalog Card
Number 83-070805
ISBN 0-9611060-0-X

CONTENTS

FIGURES

Figures Part I

Table Part II

Figures Part III

FORWARD

We would like to give special recognition to the following persons for without their patience, hard work, and understanding we doubt that the publication of this book would have been possible.

Marge Timmons, Treasurer: For her ability to find the funds to support this project.

Hope Weitzel, Sales & Production Manager: For her positive thoughts and product critique.

Darwin Nelson, Attorney: For his optimism and legal contributions.

Dominic Aspero, Vice President: The capability for patience and his artisan ability with the machining department.

Dr. Tony Castro, Armak Corporation: The old friend who devoted his valuable time and assurance.

Richard Reck, Armak Corporation: Whose faith in our overall project was so appreciated.

Mary Dunn: A lady with the needed critical eye to keep all in perspective.

A special thanks to all of our personnel at Timco for their efforts in supporting this project.

Robert Timmons, President
Timco Mfg., Inc.

PREFACE

Ground water monitoring has evolved into a science founded upon a knowledge of hydrogeologic principles and practical field experience. In addition, an understanding of the equipment and techniques available for monitoring is required. The purpose of this book is to provide such information for use in the vadose and saturated zones. The inclusion of techniques for the vadose zone underscores the importance of this system in any ground water monitoring effort. No longer will a simple monitoring well suffice in providing the detailed information required for understanding the contamination of our ground water resources.

A degree of discrimination was excercised in selecting technologies which were directly applicable for field use. Emphasis was placed upon a presentation of field proven methods which have been documented; many unproven, but highly promising, techniques were omitted. Laboratory methods which relied upon the collection of a soil sample were excluded except for a brief discussion of those methods required for the calibration or verification of a particular instrument.

The review and presentation of any emerging technology quickly becomes obsolete due to the rapid advancement of either new applications of established approaches or totally new concepts; ground water and vadose zone monitoring is no exception. The principles of the various techniques were therefore addressed on the assumption that the book will be useful even after a particular instrument becomes obsolete. The liberal use of reference material was designed to provide sources for additional detailed examination of these basic principles and to serve as a base for further study. The adage "one picture is worth a thousand words" was observed with the hope that verbosity would be avoided.

Robert D. Morrison
Newport Beach, Calif.

ACKNOWLEDGEMENTS

Individuals who have contributed to this book through correspondence or technical assistance on specific equipment and who are gratefully acknowledged are: Dr. Marios Sophocleous, Kansas Geological Survey; Professor Daniel Hillel, University of Massachusetts at Amherst; Dr. Robert Layman, Dartmouth College; Dr. James Rhoades, U.S. Salinity Laboratory at Riverside, California; Dr. William Harrison, University of Alaska at Fairbanks; Dr. George Matzkanin, Southwest Research Institute; Dr. A. Peck, CSIRO Institute of Earth Resources, Australia; Professor David Daniel, University of Texas at Austin; and Alfred Knol, Eijkelkamp, the Netherlands. Special thanks to Donna Oba for her editing and manuscript preparation. Our appreciation is extended to those individuals and organizations who have granted permission for the use of illustrations and referenced text material. Last, but not least, a special word of thanks to the entire staff of Timco Mfg., Inc. for their assistance in all aspects of this effort.

PART I
MONITORING IN THE VADOSE ZONE

A variety of techniques are available for acquiring vadose zone information. The approaches may be divided into five categories:

- soil moisture potential,
- soil moisture content,
- soil salinity,
- temperature, and
- soil pore water sampling.

Details of the first four soil water properties and their relationships are discussed in the literature.[1-4]

A. SOIL MOISTURE POTENTIAL

Soil moisture potential is a unit volume expression used to quantify the energy status of a soil water system. The total soil moisture potential includes osmotic and matric forces. Osmotic forces are associated with the dissolved ion component of the soil moisture potential while matric forces constitute those properties related to surface water tension, molecular cohesion of the water molecules, molecular water adhesion to the soil surface, and electrical forces. The soil moisture potential is inversely related to the water content of a soil; it is this property that is used for monitoring purposes.[5]

1. Laboratory Methods

Field equipment used to determine soil moisture potential usually requires the development of a soil water characteristic curve. Several laboratory methods are available for portraying this curve which is described by the soil water suction and soil moisture content. Common procedures include the modified Haines method, variations of the hanging column method, tension plate methods, and the pressure plate or membrane extractor.

The modified Haines method may be used to obtain soil water characteristic curves for suctions of 800 cm of water. The procedure requires the saturation of individual soil core samples from which excess water is extracted. When core equilibrium for a particular suction is reached, the sample is weighed. This process is repeated for several suction values. At the conclusion of the test, the sample is dried and weighed. The various sample weights are converted into water content values (percentage by volume) and graphed versus suction values. Figure 1.1 illustrates the laboratory apparatus used for a single core sample.

A similar approach is the modified Topp and Zebchuk hanging column method which allows concurrent testing of several large diameter soil cores using a reservoir of glass beads connected to the hanging water column.[7] Suction is applied through the reservoir via the hanging column. The intermediate and final sample weighing steps are similar to those of the modified Haines procedure.

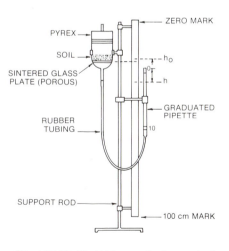

Figure 1.1 Modified Haines method apparatus.[6] (reproduced from IRRIGATION OF AGRICULTURAL LANDS, 1967, page 202, ASA Monograph No. 11, by permission of the American Society of Agronomy)

1

For suctions less than 0.8 bars tension plates are used. The maximum obtainable suction is limited to 1 bar if the core is exposed to the atmosphere and the pressure differential across the porous plate is controlled by either a vacuum or hanging water column.

A pressure plate or membrane extractor employs a gas to withdraw the soil water for a higher suction range than is possible with a tension plate. The soil core and porous plate are placed within a pressure chamber to increase the pressure differential across the porous plate. Cellulose acetate membranes can be used for suctions greater than 100 bars because of the smaller pore size of the membranes.[8] A schematic of a pressure plate and membrane extractor is shown in Figure 1.2.

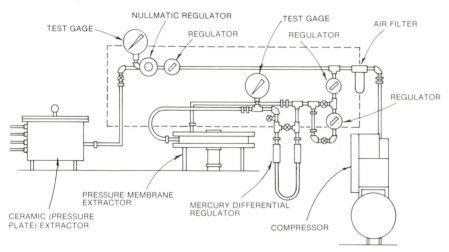

Figure 1.2 Laboratory ceramic and membrane extractor apparatus.[8] (reproduced from A FULL RANGE OF SOIL MOISTURE EXTRACTORS, 1980 by permission of Soilmoisture Equipment Corp.)

2. Field Methods

Tensiometers and thermocouple psychrometers are used to obtain soil moisture potential measurements in both the field and laboratory.

a. Tensiometers

A tensiometer (earlier terms include capillary potentiometer,[9] soil hygrometer,[10] and soil moisture meter[11]) measures the matric potential of a soil for soil suction values between 0 and 1 bar. With proper calibration the measured matric potential can be converted to the soil moisture content via a suitable soil moisture characteristic curve. The total soil moisture hydraulic potential can also be calculated with measurements from several tensiometers.

Under ideal tensiometric conditions, equilibrium between the liquid in the tensiometer and soil is achieved. Initial equilibrium between the tensiometer liquid at atmospheric pressure and the subatmospheric soil pore pressure will result in a loss of hydrostatic pressure within the vessel. This produces a vacuum that is recorded and usually expressed by the term pF.[12] As the soil pore water content increases, this vacuum is reduced by the influx of water into the tensiometer. A saturated soil will therefore theoretically record a zero value.

Tensiometer designs include porous cup and osmotic tensiometers. The basic porous cup tensiometer design consists of a fritted glass,[13] ceramic,[14] or Teflon®[15] cup attached to the bottom of a rigid plastic tube. Plastic construction is preferred because of its lower heat conduction and noncorrosive properties. The plastic tube can be double walled to reduce errors from soil temperature fluctuations. A smaller diameter copper, polyethylene, glass, Teflon®, or nylon tubing may lead from the tensiometer below the sealed top cap to a recording device. The tensiometer vessel is usually filled with deaired water although solutions of ethylene glycol (50% ethylene glycol plus 50% water), and methanol water mixtures are also used. Because these solutions have low freezing points, the tensiometer can function in cold climates.[16-17]

A variety of measuring devices can be connected to a porous cup tensiometer. These include (1) mercury filled manometers, (2) Bourdon vacuum gages, and (3) pneumatic transducers (Figure 1.3). In the case of a mercury manometer, stresses created in the liquid filled porous cup are transmitted through the liquid column to a vertically mounted glass capillary tube containing mercury. The liquid filled tensiometer and mercury train should be continuous and de-aired. Movement of the liquid from the porous cup creates a suction which causes the mercury to rise proportionally in the capillary tube (Figure 1.3a). Mercury manometers are considered accurate to about 0.2% or better between 10 to 200 cm of water.[18] Since mercury manometer tensiometers possess a high sensitivity in the lower tension range, they are best suited for use in coarse textured soils. A variation of the mercury manometer design incorporates the manometer within the tensiometer housing rather than connecting it externally to the vessel. Figure 1.4 illustrates this unit.

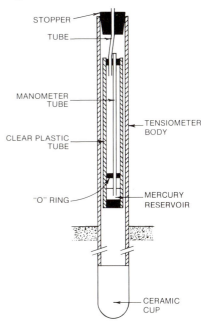

Figure 1.4 A modified mercury manometer tensiometer.[19] (reproduced from SOIL SCIENCE SOCIETY OF AMERICA JOURNAL, Volume 41, 1977, page 660, by permission of the Soil Science Society of America)

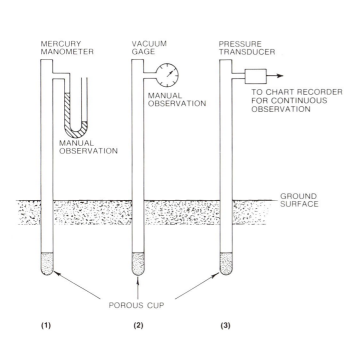

Figure 1.3 Three types of tensiometers.

Bourdon vacuum gages may be attached directly to the tensiometer (Figure 1.5) or connected via tubing.[20] Accuracy is approximately one percent of the full range of measurement. Most commercial gages are calibrated in centibars; procedures for setting the zero scale are outlined in the literature.[21,22]

Measurements are commonly made with transducers since they provide a fast response, low volume displacement system capable of continuous electrical (millivolt) measurements. One diaphragm type transducer[23,24] design is based upon the dynamic null principle whereby external pore pressure is balanced by equal internal gas pressure. In this system, no exchange of water is required to record changes in soil water suction.[25-27] Another tensiometer design relies upon a strain gage pressure transducer to measure soil suction through the deflection of a stainless steel diaphragm.[28] As soil suction changes, the tension of the wire attached to the strain gage increases or decreases; a sensitivity on the order of 2 to 3 mm of water is possible with this arrangement. A transducer with a sensing diaphragm and a linear variable differential transformer (LVDT) has also been developed to record a diaphragm deflection.[29] A transducer may even be used to measure suction changes recorded by a mercury filled manometer.[30]

Since one transducer per tensiometer is required with any of these arrangements, alternate approaches have been developed to reduce the number of transducers. One approach involves a hydraulic scanning system in which a number of tensiometers are connected to one transducer through a hydraulic scanning valve.[31-33] Figure 1.6 illustrates one such network. Another design relies upon a 24 way fluid wafer switch to line 22 tensiometers sequentially to a transducer. The latter system is insensitive to interference from temperature fluctuations, which is a problem with the scanning valve.[34]

A number of installation techniques for porous cup tensiometers have been proposed. In one procedure, a hole is augered and a volume of water equal to several vertical feet is poured into the hole and allowed to percolate into the soil. This step ensures that the immediate soil surrounding the tensiometer is initially saturated. A slurry of indigenous soil or sand and water is then placed in the hole. The tensiometer is positioned in this slurry so that the porous cup is centered at the level where measurements are to be obtained. Intimate contact between the cup and slurry is mandatory. The hole is backfilled and a bentonite seal placed at the surface.

Another installation technique consists of augering an oversized hole approximately 8 cm above the point at which measurements are to be collected. A soil auger is then used to produce a hole 8 cm deep with a diameter identical to that of the porous cup. The tensiometer is inserted into this smaller hole and backfilled with the augered soil. A bentonite surface seal completes the installation.

Once placement of the unit is completed, several operational considerations should be observed. The greatest source of error is from air entry into the vessel via the porous cup. Since diffusion of air into the unit occurs even under optimum conditions via dissolved air in the water entering the vessel,

Figure 1.5 Bourdon gage tensiometer. (photograph ©1983, Timco Mfg., Inc., Prairie du Sac, WI)

a deaired solution is recommended to minimize equalization of the internal unit pressure with the atmosphere.[35] Many tensiometers incorporate a deairing feature into their design or a modification so that a deairing system can be attached. When soil suction exceeds 0.8 bars for longer than one day, a large volume of the solution is extracted from the tensiometer rendering the unit ineffective. The tensiometer must then be flushed out to remove air bubbles and refilled with a deaired liquid: air in the unit may also be removed via vacuum.[36,37] After the unit is deaired, several hours are necessary for the unit to re-establish equilibrium prior to subsequent measurements.

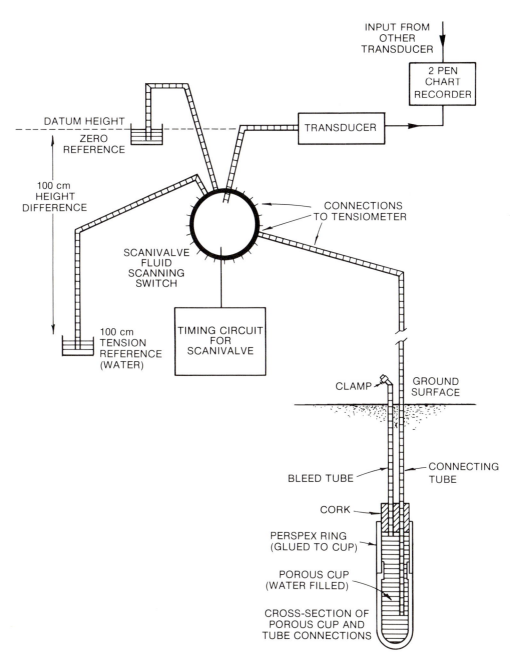

Figure 1.6 A scanning valve tensiometer system.[32] (reproduced from JOURNAL OF HYDROLOGY, Volume 33, 1977, by permission of the Elsevier Scientific Publishing Co., Amsterdam, Netherlands)

An osmotic tensiometer[38] uses a confined aqueous solution (polyethylene glycol), rather than deaired water as the reference solution (Figure 1.7). A membrane separates the confined solution from the soil pore water though small ions and molecules can pass through the membrane.[39,40] Once equilibrium between the reference solution and the soil water is attained via the permeable membrane, a pressure transducer is used to measure the subsequent soil moisture changes. Osmotic tensiometers, however, appear to be susceptible to fluid leakage and associated operational difficulties which limit their use.

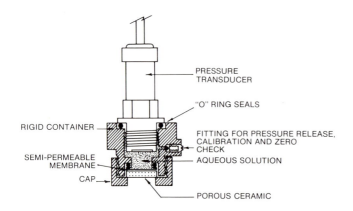

Figure 1.7 Osmotic tensiometer.[38]

Installation of an osmotic tensiometer is performed by augering to the necessary depth, placing the unit in the hole, and surrounding the sensor with a slurry of fine sand and water. Firm contact with the membrane is critical. A bentonite seal is placed above the unit.

With the installation of several tensiometers at various depths, the hydraulic gradient can be calculated according to the following relationship:[41]

$$i = \frac{\left[(S_{N+1} + d_{N+1}) - (S_N + d_N)\right]}{d_{N+1} - d_N} \tag{1.1}$$

where
$S_1, S_2, S_3, ..., S_n =$ matric suction values in centimeters of head (millibars),

$d_1, d_2, d_3, ..., d_n =$ tensiometer depths (cm), and

$i =$ average hydraulic gradient between depths d_n and d_{n+1}.

This relationship assumes that the hydraulic gradient is vertical; where three dimensional gradients are required, additional tensiometers are installed. Streamlines may be estimated by interpolating the (i) values along the hydraulic gradients (i.e., the difference in hydraulic head

6

between two points divided by the distance between the points). When (i) is positive, flow is downward and unsaturated. When (i) is negative, the flow is upward with the millibar value equivalent to the distance of the porous cup below the piezometric surface.[42] Automated tensiometer systems have been designed to measure either the piezometric surface [negative (i) value], or soil moisture tension [positive (i) value], depending on the soil water status.[43]

When used for measuring the hydraulic gradient in a saturated soil, measurements are comparable to those obtained with a monitoring well with one advantage: equipotential lines above and below the water table can be determined.[44] One and two dimensional flow models can therefore be validated with this information.[45] A similar approach may be used for developing the hydraulic gradient within a snowpack.[46,47]

Tensiometers provide a simple means of obtaining matric potential data. Information on the soil moisture for saturated and unsaturated conditions can be collected rapidly, especially in the case of a transducer tensiometer system. Tensiometry, however, provides an indirect measurement of soil moisture and the resulting data should be viewed in this context. Tensiometric data may also represent a number of hysteretic points; the researcher must therefore know whether the reading was obtained during a wetting, drying, or intermediate (scanning curve), period. Diurnal variations in temperature related vapor pressure gradients and distillation transfer between the cup and soil are other potential sources of error. Air entry into the vessel can result in measurement errors or unit failure.

Tensiometers are best suited for use in coarse, sandy soils where over 90% of the soil moisture range can be observed. Installation in a clayey soil results in less satisfactory data as a significant change in suction corresponds to a minor volumetric change in water content.

b. Thermocouple Psychrometers

A psychrometer measures the relative humidity of a system. In soil applications they are usually restricted to suctions between about 0.9 to 72 bars which is beyond the range of tensiometers. Psychrometers are, therefore, ideal for vadose zone monitoring in arid regions.

Psychrometry measurements are classified as either hygrometric (dew point) or psychrometric (wet bulb). Thermocouple psychrometry is of the latter variety and is based upon the measurement between a non-evaporating (dry bulb) and evaporating (wet bulb) surface. The temperature difference between these surfaces constitutes the relative humidity, which may be calibrated to a soil water potential value. All psychrometers function according to this principle but differ by how water is applied to the evaporating surface. This distinction provides the basis for categorizing thermocouple psychrometer designs as belonging to either the Spanner or Richards and Ogata variety.

The Spanner psychrometer measures the condensation of atmospheric moisture within the psychrometer chamber on an evaporating thermocouple (B in Figure 1.8b). During the brief period of thermocouple cooling produced by the application of an electromotive force (emf) at the wire juncture, a minute quantity of water is condensed.[49] The temperature difference between the reference electrode and thermocouple (e.g., the dry and wet bulb) during this evaporation period is recorded and calibrated to obtain a soil moisture potential value. The Richards and Ogata design uses a small silver ring supported by two wires which constitutes the evaporating juncture.[50,51] The steady state temperature change during the evaporation of the water droplet at the wire loop is subsequently recorded and correlated to obtain the soil moisture potential.[52]

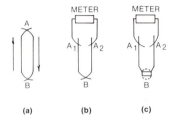

Figure 1.8 Thermoelectric effects used in thermocouple psychrometry. **(a)** The Seebeck effect; current flows due to a temperature difference between A and B. **(b)** Measurement to temperature difference between A and B. B may initially be cooled by the Peltier effect (Spanner psychrometer). **(c)** Maintenance of permanently wet junction at B. (Richardson and Ogata psychrometer)[48] (reproduced from H. Barrs, Determination of Water Deficits in Plant Tissues. T. Kozlowski (Ed.), WATER DEFICITS AND PLANT GROWTH, © 1968, Academic Press, N.Y., NY)

The majority of psychrometer designs are of the Spanner variety. Literature describing psychrometer design criteria and construction techniques therefore deal primarily with the Spanner psychrometer.[53-61] Basic components of this psychrometer are a thermocouple, a reference electrode, a heat sink, a porous ceramic bulb or wire mesh screen, and a recorder (Figure 1.9). The thermocouple consists of two wires of dissimilar metals. Chromel-P and constantan are often used because of their sensitivity and high Peltier coefficient. Bismuth and bismuth with 5% tin may also be used to extend the useful lower limits of measurement provided with Chromel-P or constantan. A wire diameter of 0.0025 cm is generally used since it results in a minimum thermal and vapor disturbance in the psychrometer chamber.

The reference electrode is often embedded in a solid portion of the psychrometer. A copper constantan thermocouple of 24 gage or smaller is used for an accuracy to the nearest °C.[63] Copper wires often serve as psychrometer heat sinks although thermal gradients along the wire may introduce erroneous large voltage measurements. Some removable units have used a gram of copper or brass located near the reference junction as a heat sink.[64]

The porous bulb or screen component provides protection for the thermocouple, maintains a fixed void within the soil, and permits rapid vapor pressure equilibrium between the soil and the thermocouple chamber. The material and geometry of this component can have a pronounced effect

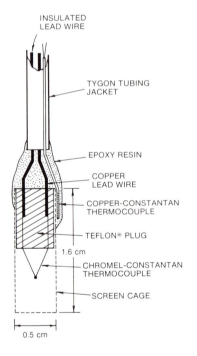

Figure 1.9 Spanner psychrometer.[62]

8

upon the accuracy of the measurements. Ceramic cups[65,66] and a 200 Dutch stainless steel screen weave[67] are the two most commonly used materials. The time required for vapor equilibrium is directly affected by the material. Figure 1.10 summarizes the results of one study in which unprotected (open end window), screened, and ceramic cups were evaluated.[60] The response time lag indicated for ceramic in Figure 1.10 can be reduced by wetting the ceramic prior to installation. For soils with a high shrink swell capacity, the wire screen enclosure is preferred to the open or ceramic design.

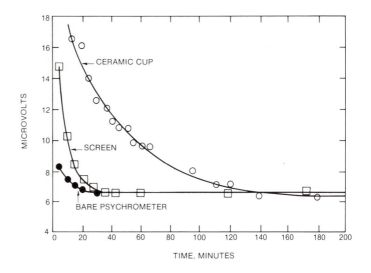

Figure 1.10 Lag in response time of a thermocouple psychrometer to vapor equilibrium over 0.3 molal KCl solution at 25°C.[60]

The geometry of the cup or screen will affect unit performance due primarily to temperature gradient errors within the chamber. A cylindrical ceramic unit with a centrally located thermocouple minimizes this effect while open end window geometries produce the greatest measurement errors. The placement of a non-evaporating surface plug in the end of the ceramic unit[68] and chamber reduction minimizes this effect.

A variety of automatic and semiautomatic recorders are available for psychrometer measurement.[69,70] A potentiometer or galvanometer is often used for direct current measurement. A microvolt potentiometer designed for this purpose consists of a wall galvanometer or a null detector with manual counterbalancing of the thermocouple output. An automated potentiometer coupled with an electronic amplifier has also been used.[71] An automatic system composed of a scanner, capacitor, voltmeter, and recorder has been fabricated for use with a Spanner psychrometer.[72] A stepping relay added to the arrangement allows 150 psychrometers to be read in one hour. Most galvanometers use the ballistic method in which the maximum deflection is measured[73,74] for either the Spanner or Richards and Ogata design. The use of a potentiometer or galvanometer requires a sensitivity of ± 0.01 μv in the usual measuring range of 0 to 30 μv.

9

Modifications to the basic Spanner design include a temperature compensating double loop psychrometer, and a psychrometer and salinity sensor combination. The double loop psychrometer (Figure 1.11) was developed to reduce the significant error which occurs with an abrupt temperature change by incorporating a built in temperature compensator. When the thermocouples are connected with opposite polarity, most of the extraneous current from temperature fluctuations are eliminated.[75-77]

A thermocouple psychrometer and salinity sensor have been combined to provide simultaneous matric and osmotic potential measurements (Figure 1.12).[78] Osmotic readings accurate to within ±10% are obtainable with the salinity sensor; the matric potential is inferred from the difference between the water potential and osmotic potential.

Other improvements in the Spanner psychrometer include increasing the wire diameters,[79,80] heavier Peltier currents, and longer cooling times. These changes are designed to allow increased discrimination between osmotic and matric potentials since changes in one component can be masked by a flux of the other in the opposite direction when accompanied by temperature changes.[81]

Thermocouple psychrometry is based upon the relation between water potential and the relative humidity on an evaporating surface at any given temperature (T):

$$\psi = \frac{RT}{V} \ln \frac{p}{p_0} \qquad (1.2)$$

where
ψ = water potential,
R = ideal gas constant for water vapor,
T = absolute temperature,
V = molar volume of water, and
$\frac{p}{p_0}$ = ratio of actual (p) to saturation vapor pressure (p_0) (i.e., relative humidity).

This relation assumes that water vapor acts as an ideal gas (i.e., the error introduced for corresponding pressure is negligible). Sensitivity therefore decreases with increasing pressure and increases with higher temperature. Theoretical and experimental work has confirmed that for any temperature or pressure, relative humidity measurements within a stabilized chamber can be derived from a psychrometer calibrated at any other temperature or pressure resulting in a fairly linear scale.

Calibration curves for porous bulb psychrometers are developed by immersing the unit in a standard solution. KCl or NaCl solutions of 0.1, 0.3, 0.5, 0.8, and 1.0 molar are usually used.[82-84] The flask containing the calibration solution and psychrometer is placed in a constant temperature bath, and the psychrometer output is recorded after temperature equilibrium is attained.[85] The bath temperature can then be changed or the flask transferred to another standard solution bath. Calibration curves for psychrometers encased by a wire screen are developed by suspending the psychrometer over NaCl standard solutions in sealed chambers under isothermal conditions.[86] Once calibration is completed the entire unit should be thoroughly cleaned. The housing can be placed in an ultrasonic cleaner with a 50% ethanol solution

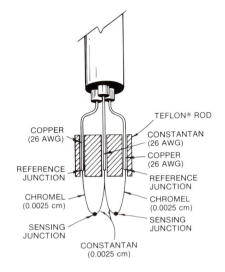

Figure 1.11 Double loop, temperature compensating psychrometer.[70]

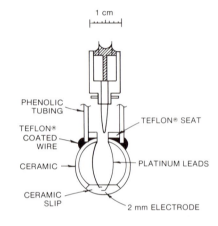

Figure 1.12 Combined thermocouple psychrometer and salinity sensor.[78] (reproduced from SOIL SCIENCE SOCIETY OF AMERICA PROCEEDINGS, Volume 34, 1970, page 570, by permission of the Soil Science Society of America)

followed by rinsing in distilled water. Thermocouple cleaning may be performed by rinsing in a high grade acetone followed by drying with an air gun. The process is repeated using cool, distilled water.

Field installation for shallow monitoring is often performed by placing the psychrometers in the sidewalls of a trench. The sensors should be placed horizontally to minimize temperature gradient effects; lead wires should also be placed horizontally several feet before being terminated at the surface. For deeper installations, a perforated polyvinyl chloride (PVC) well point with the psychrometer placed within the tube is used.[87] This arrangement allows the removal, recalibration, and reinsertion of the psychrometer without damage to the unit.

Thermocouple psychrometers provide a wide range of water potential measurement versatility, especially in dry soils. Psychrometers may be used to determine the water potential gradient as a function of depth;[88] preliminary evidence suggests that they may also be employed in snowpack studies.[89] The major limitation of a psychrometer is related to a lack of thermal or vapor control within the psychrometer chamber[90] and contamination of the thermocouple. If a temperature gradient exists within the chamber, the reference junction temperature may be quite different from the sample temperature, thereby resulting in measurement error. A similar error will occur if a thermal gradient develops in the chamber during measurement.

Contamination of the chamber interior or thermocouple can result in erroneous readings. Foreign substances within the chamber may act as a vapor sink, thereby altering the humidity within the chamber. Ceramic is especially susceptible to subsequent psychrometer contamination from KCl or NaCl standards which remain in the pores and eventually migrate to the thermocouple. Deposition or corrosion of the thermocouple will result in erroneous measurements. As the thermocouple is cooled, the contaminated substance dissolves in the water; as more water is condensed, the solution becomes diluted. The measurement will therefore be lower than the actual value since the vapor pressure of the solution is lower than for pure water. Introduced contamination sources usually originate from the manufacturing process, calibration, installation, or from elements in the surrounding soil (Figure 1.13).

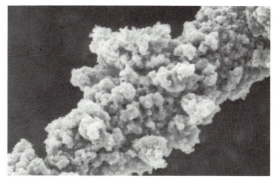

(a) Sulfur-rich corrosion products on a constantan thermocouple psychrometer wire.

(b) Corrosion products (rich in aluminum, silicon, and iron) on a constantan thermocouple psychrometer wire.

Figure 1.13 Photomicrographs of constantan wire corrosion from acidic soils.[91] (reprinted with permission from the American Society for Testing and Materials, 1916 Race Street, Philadelphia, PA 19103. STP 746)

B. SOIL MOISTURE CONTENT

The soil moisture content is the ratio of the weight of the water to the weight of solid particles in a soil mass. The moisture content expressed on a volumetric basis is a critical item to be defined for vadose zone monitoring.

1. Laboratory Methods

Laboratory methods for determining the moisture content of a soil are useful for verification or calibration of many types of in situ field instruments. The five basic testing approaches are thermo-gravimetric, chemical extraction, mechanical extraction, immersion, and penetration. Detailed procedures and discussions of these techniques are available in the literature.[92-96]

Gravimetric methods consist of weighing a wet soil sample followed by oven drying at 105°C until a constant weight is attained. Microwave, infrared, and vacuum drying are also used. The weight of the evaporated water is expressed as a percentage of the dry soil weight. The volumetric water content is described as:

$$\theta = \frac{W_w \, Y_d}{W_d \, Y_w} \ 100\% \tag{1.3}$$

where θ = volumetric water content (%),

W_w = weight of water (g),

W_d = dry weight of soil (g),

Y_d = oven-dry bulk density (g/cm^3), and

Y_w = water density (g/cm^3).

A number of variations to this method include freeze drying (lyophilization), distillation, heating in oil,[97] dessicant weight gain, and alcohol burning.[98,99]

The three types of chemical extraction are alcohol, calcium carbide (hydride), and the Fischer approach. In the alcohol extraction method, the moisture content is determined by the density (hydrometer method) of the alcohol and water mixture after extraction from the soil sample. The calcium carbide method is usually related to the decrease in the weight of the carbide mixture after evolution of acetylene and the rise in pressure in a closed vessel which contains the mixture that measures the volume of gas produced.[100] The calcium carbide reaction with the soil moisture proceeds as:

$$CaC_2 + 2H_2O \longrightarrow Ca(OH)_2 + C_2H_2 \tag{1.4}$$

The pressure of the C_2H_2 gas produced within the vessel is correlated with the moisture content. A similar procedure is employed in the hydride reaction:

$$CaH_2 + 2H_2O \longrightarrow Ca(OH)_2 + 2H_2 \tag{1.5}$$

12

In the Fischer method, the soil sample is dissolved or leached with a solvent followed by titration with an iodine sulfur dioxide and pyridine in methanol solution.[101-102] Determination of the titrated end point is related to the soil moisture content.

Mechanical extraction of soils with a high moisture content is possible with pneumatic or hydraulic presses. The weight before and after compression is used to calculate the water content.

Immersion (also pycnometer or displacement) techniques measure the change in the specific gravity of the soil sample by various liquids.[103,104] Water, alcohol, and alcohol acetone salt solutions have been employed. One technique relies upon the change in soil conductivity created with the displacing fluid. This value is corrected for temperature and correlated with the soil moisture content.[105]

2. Field Methods

Indirect field methods used to determine soil moisture content include electromagnetic, electrothermal, and nuclear techniques.

a. Electromagnetic Methods

Electromagnetic techniques include those methods which rely upon the effect moisture has upon a soil's electrical properties. The general relationship between soil moisture and the frequency dependent dielectric response function is described by:[106]

$$\xi(\omega) = \xi_r(\omega) + j\xi_i(\omega) \tag{1.6}$$

where $\xi(\omega)$ = dielectric response function,

   $\xi_r(\omega)$ = the real part of ξ (e.g., the energy stored by dipoles aligned in an applied electromagnetic field),

   $\xi_i(\omega)$ = the imaginary part of ξ (e.g., the energy dissipation rate in a medium; in some soils, an ionic conductivity term is used with $\xi_i(\omega)$),

   j = the square root of -1, and

   ω = angular frequency.

This relationship allows soil moisture measurements based upon resistivity (ξ_i), capacitance (ξ_r), or both. Three methods which depend upon this characteristic include moisture blocks, the four electrode approach, and capacitance sensors.

(i) Soil moisture blocks

Soil moisture blocks are sensors which measure soil moisture as a function of the resistance between two electrodes embedded in a porous material that is in soil equilibrium. Since the resistance is actually a measurement of the soil matric potential; soil salinity, texture, temperature, and minor reactance effects are presumed to be constant.[107]

The basic elements of a soil moisture block are two electrodes, a matrix which encapsulate the electrode, and a recorder (Figure 1.14). Electrodes have been fabricated from palladium, Monel, copper, magnesium, brass, stainless steel, pure nickel, and sterling.[108,109] Electrode geometries

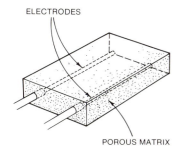

ELECTRODES

POROUS MATRIX

Figure 1.14 Rectangular soil moisture block.

13

include single wires, chrome plated brass tips, and screens placed in parallel or concentric arrangements within the porous media. A screen type electrode consisting of 20 mesh stainless steel appears to perform better than single wire electrodes due to improved contact with the matrix.[110] A concentric electrode design minimizes stray electrical currents from the sensor; hence, many moisture blocks are cylindrical.[111-114] Multiple electrodes are often included in one sensor to eliminate erratic contact resistance which commonly occurs with a two electrode sensor.[115]

Gypsum,[116,117] ceramic,[118] Castone (dental stone powder),[119] fiberglass,[120] and nylon[121] have been used for the porous matrix. Gypsum (plaster of paris or Bouyoucos) blocks are usually combined in an 80 parts water to 100 parts gypsum mixture.

This mixture provides a buffer against salt concentration by maintaining an electrolyte concentration of about 2200 to 2400 ppm. Excessive salt concentrations ($\geq$ 5000 ppm) will result in significant interference; measurement deviations, however, have been detected in concentrations of as little as 1800 ppm.[122] Although the gypsum provides a certain degree of buffering capacity, dissolution of the material results in limited field use; in well drained soils the block may be operational up to 7 years while no more than 1 or 2 years can be expected for blocks placed in wet or saturated soils. In order to reduce block dissolution, nylon, fiberglass, and resin are often added. Combinations of gypsum with other materials also extends the range of the block since gypsum, with a pore space percentage of about 44%, is responsive primarily at matric potentials greater than 0.3 bar tension.[123] A nylon gypsum block is most sensitive below 0.3 bar, while a fiberglass gypsum combination with large electrodes provides the best combination for readings between 0 and 15 bars.

Ceramic and Castone are stable materials that are not significantly altered with time unless a biofilm accumulates on the material.[124] These materials, however, do not provide any buffering capacity and are therefore highly sensitive to variations in soil salinity.

Fiberglass provides a durable matrix although it is susceptible to decomposition in many soil environments. Fiberglass has also demonstrated a chemical reaction or ion exchange reaction with soil salts. Fiberglass blocks reach soil moisture equilibrium faster than gypsum blocks but offer no buffering capacity against salt interference.[125]

Nylon is a highly durable matrix with an in situ life span greater than 5 years even in saturated conditions. Nylon blocks provide a rapid response to moisture changes, thereby resulting in the rapid diffusion of salts which are a major factor especially at low moisture values. The thin matrix also minimizes a small variation in moisture content between the outer and inner portions of the sensor, which occurs when evaporation on the block surface is greater than water movement within the block. Nylon offers no buffering capacity, however, and is usually fabricated with gypsum.[126]

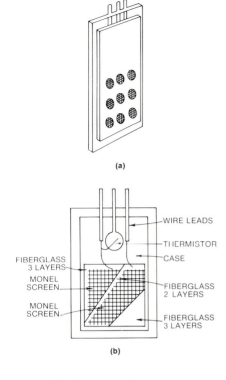

(a)

(b)

Figure 1.15 Fiberglass and Monel soil moisture sensor with thermistor.[120] (reproduced from SOIL SCIENCE, Volume 67, 1948, page 426, ©1948, Williams and Wilkens Co.)

Sensor size and geometry affect the time required to attain equilibrium: small sensors generally reach equilibrium faster than larger ones. Sensor geometries include rectangular and cylindrical shapes, tapered probes, and a variety of multielectrode configurations. An example of a rectangular block (shown in Figure 1.15) consists of a Monel screen (electrodes), a fiberglass cloth sandwich, and a nonlinear temperature sensing element (thermistor). A multielectrode probe containing three parallel stainless steel electrodes separated by cast gypsum is illustrated in Figure 1.16.[127,128] Another multielectrode probe design employs three concentrically placed Monel mesh electrodes to minimize the effect of stray currents.[129] The inner electrode is wrapped with fiberglass while the annulus between the three electrodes is filled with gypsum. The inner and outer electrodes provide measurements over a wide range of matric potentials; the middle and outer electrodes are used for measurements at higher matric potentials. A thermistor is placed in the center of the innermost electrode so that temperature corrected resistance values can be calculated.

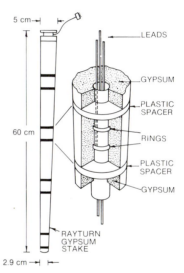

Figure 1.16 Multielectrode probe.[128] (reproduced from SOIL SCIENCE, Volume 86, 1958, ©1958, Williams and Wilkens Co.)

Moisture block measurements are usually obtained with a moisture bridge based upon the Wheatstone principle. Since polarization and electrolysis occur with a direct current, an a.c. meter is used. The capacibility to record resistance values from about 50 to 1000 ohms with the meter will suffice for most cases, although meters with larger ranges are common. Moisture bridges have been modified so that continuous readings of electrical resistance and soil temperature are obtained from numerous readings.[130,131]

Inconsistencies in the construction of soil moisture blocks require that either individual calibration curves be developed for each sensor or a block which is considered representative of a number of units be selected for curve development. In the latter approach, the blocks are placed in distilled water and the resistance measured; units varying more than 50 ohms are not included. Calibration curves are preferably presented as a function of resistance versus soil water suction.

Soil suction can be used to develop curves of water suction versus soil moisture content by placing the blocks in a representative soil sample and measuring the extraction of the soil pore water via a pressure plate apparatus. Other approaches include calibration in distilled water[132] and in representative soils placed within a wire basket.[133,134] In the latter method, the resistance block is placed in a cloth lined wire basket surrounded by a 1.2 cm layer of soil. The soil is wetted and the unit placed in a humidity chamber to drain; cycles of five hours of laboratory evaporation and 19 hours in the humidity chamber are alternated until the soil no longer loses moisture in the evaporation cycle. Soil weights are recorded and resistance readings obtained during removal from the humidity chamber. These measurements provide a calibration curve. For precise values, resistance should be converted to the resistance for an identical moisture condition at a standard temperature and block material.

In arid regions in soils with a high salt content, moisture blocks are often calibrated in situ with gravimetric verification to minimize the resistance caused by a high salt concentration. For work in humid areas where salt contents are lower, in situ calibration is unnecessary. This type of calibration may result in greater curve errors than caused by the potential effect of salt interference.

Moisture blocks are saturated in distilled water prior to horizontal installation in a borehole or sidewall. For individual placement, a slurry of 200 mesh sand is poured into the bottom of the hole. The block is forced into this slurry and backfilled with the excavated soil in the order in which it was removed. In multiple installations this procedure is repeated at the various depths.

Single or multiple sensor installation in a trench or borehole side wall is preferred to other methods due to its minimum disturbance of the overlying soil. In either a single or multiple array, a hole with the approximate dimensions of the moisture block is made in the side wall. The excavated soil from the side wall is retained. The moisture block is placed into the hole and the retrieved soil from the side wall is used to fill any void between the sensor and the soil. The borehole or trench is backfilled and compacted to either the next installation or the surface.

Measurements should not be recorded for at least 24 hours after installation to allow for sensor equilibrium with the soil. The same moisture meter used for calibration should be used for all field readings. Once the lead wires are connected to the meter terminals, one to two minutes are recommended for unit stabilization before recording resistance values.

Moisture blocks are an inexpensive tool for soil moisture monitoring due to the low cost of the individual sensors. The ability to record measurements from many units over a large area via an automated recording system is another advantage. Problems include sensitivity to changes in salt concentration, temperature effects, sensor integrity, proper soil contact, limited operational ranges of many sensors, and pronounced hysteresis[135] during calibration and field use.

16

(ii) Four electrode method

Electrical conductivity or resistivity can be measured with electrodes in the soil. The primary factors governing the conductivity or resistivity values between electrodes are the dissolved electrolyte concentration, temperature, bulk density, and moisture content: the quantification of the first three factors allows the determination of the fourth factor.

The two most common four electrode arrangements are (1) four electrodes (e.g., two current electrodes and two potential electrodes) protruding from the ground surface (Figure 1.17) and (2) electrode burial at prescribed electrode spacing intervals.[136,137] In the first arrangement, an electrical current is passed through the two outer electrodes and measured along with the potential drop across the two inner electrodes. Surface electrode configurations usually consist of individual placement or stationary electrode positioning on a metal rod.[138] An electrode spacing of 7.6 cm between the fixed electrodes has been used although smaller intervals are common.

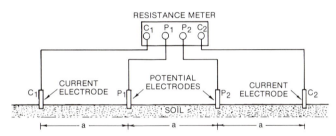

Figure 1.17 Wenner array with a=interelectrode spacing; C_1 and C_2, the current electrodes; and P_1 and P_2, the potential electrodes.

Electrodes are also buried rather than placed on the soil surface although the principles are identical. One arrangement consists of metal bands buried between various soil layers.[139] A similar configuration relies upon the conductance between two parallel plates separated into 40 individual metallic strips. Copper strips, 150 mm long by 1 mm wide, are encased in a plastic plate from which they protrude 2 mm. The forty bands represent a separate, measurable circuit on each of the two electrodes.[140]

A calibration curve is necessary for determining the soil moisture content from a four electrode resistance or conductance measurement. The apparent resistance or conductance, soil salinity, bulk density, temperature, and electrode spacing may be correlated by calibration of the electrical potential with the moisture content of the soil core. Readings from electrodes placed on the surface or buried in a dry soil column are recorded throughout a wetting cycle. Soil samples are collected during the test and the moisture content is verified gravimetrically. A curve correlating the electrical resistance reading and soil moisture content is then prepared.

The shape of the resulting calibration curve is affected by the dissolved electrolyte concentration and cross section of the solution filled pore space. Temperature corrections are developed with calibration charts obtained at various soil temperatures. A 17°C change in soil temperature may introduce a 1% error in the specific conductivity. Hysteresis does not represent a major influence on the calibration curve shape.

Intimate contact between the electrodes and soil is required for representative readings. Substantial error can result from either electrode displacement or inaccurate electrode spacing. Thermocouples for temperature measurements are often installed with the buried electrodes or at the same depth.

Four electrode methods provide an indirect method of expediently obtaining soil moisture values. Once the appropriate calibrations have been performed, a near continuous monitoring of an event is possible with little (buried electrodes) or no (surface electrodes) soil disturbance. When used in conjunction with another more direct technique, the method provides an optimum combination for soil moisture determination. Large errors can occur, however, by increasing the area of concern (e.g., via electrode spacing), monitoring in anisotropic soils, poor electrode soil contact, spatial variability of the dissolved electrolyte concentration, and monitoring in soils subject to freezing.

(iii) Capacitance sensors

Capacitance (dielectric) sensors are based upon the relationship between the soil moisture content and the capacitance of the soil measured by a pair of embedded electrodes. Capacitance sensors incorporate the coil as the dielectric (i.e., part of the measurement circuit) as opposed to methods which absorb moisture into a resistive or dielectric element.

A fundamental electrical property of a material is its dielectric permittivity (ϵ) which is proportional to the dielectric constant (k) by:

$$\epsilon = k\epsilon_0 \qquad \text{(1.7)}$$

where ϵ_0 = permittivity of a vacuum (8.85×10^{-12}/ Farads/meter in the mks system of units).

Since ϵ and ϵ_0 have the same dimensions, k is dimensionless. The relationship between the dielectric constant and the capacitance (C) of a sensor consisting of two electrodes embedded in a soil is:[141]

$$C = \alpha \epsilon_0 k \qquad \text{(1.8)}$$

where   k = the dielectric constant,

α = electrode geometry constant, and

ϵ_0 = permittivity of the vacuum.

For most solid soil components, the dielectric constant ranges from about 2 to 4 while the dielectric constant for free water is approximately 79 between 15 and 35°C. The dielectric constant varies primarily with the number of water molecules present per unit volume of soil (usually expressed in cm³ or grams per cm³ of soil) in the zone of influence between the electrodes. The resonant circuit containing the electrodes and moist soil will therefore oscillate at a frequency dependent on the soil moisture content.

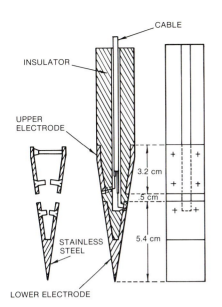

Most capacitance sensors incorporate an oscillator circuit within the sensor. A coaxial cable connects the sensor to an external power supply and frequency monitoring device. A variety of sensor geometries and circuitries are available with this basic arrangement to accomodate the various monitoring applications, pore water characteristics, and soil properties.

While most capacitance sensors are cylindrical, other geometries include plate, probe, and rectangular shapes. A sensor shaped like a plate, for example, uses the fringing field of a low loss, dielectric slab waveguide for fine textured soils while a nearly identical unit relies upon waves launched from the dielectric slab for use in coarser materials.[142] By increasing the electrode gap and perimeter, a larger zone of influence is produced: a better moisture content average therefore results. The probe sensor (Figure 1.18) consists of two metal semicylinders whose diametral planes coincide with the electrode surfaces.[143] The tapered shapes ensure intimate contact between the probe and the soil.

Cylindrical capacitance sensors appear to be superior to other shapes. A cross section of one sensor is shown in Figure 1.19. A radius of influence of about 1.3 to 2.5 cm is measured with this design. Another cylindrical unit with the electrodes placed in a coaxial arrangement is depicted in Figure 1.20[145-147] The sensor is attached to a Wien bridge circuit for either conductance or capacitance readings.

Figure 1.18 Capacitance probe.[143] (reproduced from A. Thomas, JOURNAL OF SCIENTIFIC IN-STRUMENTS, Volume 43, pages 21-27, ©1966, with permission of The Institute of Physics)

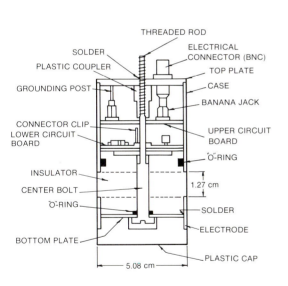

Figure 1.19 Cylindrical capacitance sensor.[144] (reproduced with permission from H. Wobschall, IEEE TRANSACTIONS GEOSCIENCE ELECTRONICS GE-16, ©1978, IEEE)

Figure 1.20 Cylindrical capacitance sensor with coaxial probe arrangement.[145]

A calibration curve is required for each capacitance sensor using a representative soil sample from the installation site.[147] To develop a curve, the sensor is inserted into a dry soil to which increasing amounts of distilled water are added. The bulk density and soil temperature should remain constant. The capacitance or frequency is recorded, and this measurement versus water content per unit volume (gravimetrically determined) is plotted.

Another technique allows the calibration of one reference sensor to which others are adjusted.[148] In this method, a calibration curve (gravimetrically verified) is developed for one sensor with the soil of interest. This reference sensor is then placed in solutions of different dielectric constants such as air, dodecane, octanol-1, butyl alcohol, ethylene glycol, and water. The other sensors are also placed in these solutions and the results plotted. Since the relationship is linear, data from any sensor can be correlated to the reference sensor.

Capacitance sensors can be permanently installed in the soil, for continuous measurement, or lowered down an access tube, for soil moisture profiling. In situ placement requires minimum soil disturbance around the sensor; contact between the sensor and soil is essential. Capacitance probes are tapered so that they can be pushed into a hole slightly smaller than the probe to ensure contact. A thick walled Shelby tube with a surface guide fixture has been used to bore a hole for the sensor shown in Figure 1.19. The soil core is removed and the sensor is placed in the hole followed by backfill compaction with a modified Harvard miniature tamper. Cylindrical units with extending electrodes (Figure 1.20) are installed by forcing the tapered probes into the undisturbed soil. The hole or trench is backfilled with indigenous soil, followed by a grout or bentonite surface seal.

Several designs allow the sensor to be lowered inside an access tube for soil moisture profiling.[149,151] The access tube should be of a material with a low relative dielectric constant. Care is required in the installation of the access tube to avoid the formation of cavities or excessive soil disturbance around the tube. This procedure assumes that the soil is homogenous and identical in bulk density to the soil used for sensor calibration.

Potential sources of measurement error are associated with soil temperature, salt concentration of the soil pore water, soil bulk density, and frequency. The influence of temperature upon measurement is proportional to the soil moisture content; as the moisture content increases so do the chances of error from temperature. Volume changes due to increased moisture can further complicate the problem. Despite the potential for error, temperature fluctuations less than ±5°C can be ignored while larger changes are corrected with temperature compensation formulas.[152,153]

The effect of electrolytes upon capacitance measurements will depend upon the concentration. Calibration curves, for example, exhibit significant changes with increased ionic concentrations.[148] Frequencies in the low megacycle range also appear to be more sensitive to ionic concentration than higher frequencies.

Variations in bulk density with time exert an indirect effect upon the relationship between the dielectric constant and soil moisture.[154-156] A reliable method for compensation of this effect, however, has not been developed. In soils where such a condition is known to exist, laboratory investigation of the soil at various densities and moisture contents is required to examine the behavior of the particular soil upon the dielectric function.

Capacitance sensors offer a reliable, simple, and accurate (±2%) means of determining soil moisture content. The linear relationship between soil moisture at selected frequencies is valid for soils of varying texture, organic matter content, temperature (≈0 to 30°C), and ionic concentration. The radius of the measurement can be controlled through proper sensor selection. Disadvantages include the potential interference from ionic concentration in the soil system, decreased sensor sensitivity with time, errors in high shrink swell soils, and problems resulting from improper installation.

b. Electrothermal Methods

Electrothermal methods are based upon the rate of thermal dissipation from a heat source through a media of low thermal conductivity. The increase in temperature at any point in the material depends upon the thermal conductivity which can be correlated with the soil moisture content.

Steady state and transient methods are used to measure thermal conductivity. Due to difficulties in obtaining valid results for soils once equilibrium conditions are reached with steady state techniques, a transient heat flow principle is used.

Three types of electrothermal (also referred to as heat diffusion and heat dissipation) transient devices are: (1) a porous block with embedded electrical elements; (2) a direct contact type with the electrical elements in contact with the soil; and (3) a modified direct contact probe or cell.[157]

The porous block design includes a miniature heater, a temperature sensor, and associated circuitry. Figure 1.21 illustrates one design consisting of a germanium P-N junction diode surrounded by a 200 cm Teflon® coated copper wire coil embedded in a cylindrical Castone or ceramic matrix.[124,158,159] Heat shrinking rings are placed in the diode case to retain heat for the heater wire. A similar unit incorporates a completely insulated thermistor encapsulated in gypsum or Castone, designed to eliminate sensitivity bias from soil salinity.[160] A modified Wheatstone bridge and galvanometer are used for measurement.[161]

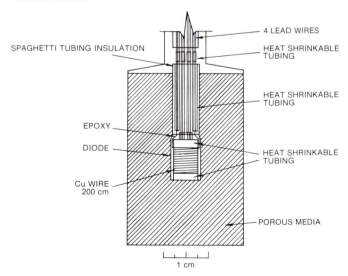

Figure 1.21 Porous block electrothermal sensor.[158] (reproduced from SOIL SCIENCE SOCIETY OF AMERICA PROCEEDINGS, Volume 35, 1971, page 226, by permission of the Soil Science Society of America)

The porosity of the block matrix should be nearly identical to that of the soil in which it is to be placed. Without a comparable capillary potential between the block and the surrounding soil, moisture equilibrium between the two materials will not occur. The resulting soil moisture measurement will therefore not reflect the true soil moisture content. The difficulty of providing a suitable matrix to correspond to a particular soil is a major limitation of porous block designs.

Direct contact electrothermal devices generally consist of a single heating element or wire wound around glass to maximize the unit's thermal conductivity.[162,163] Several designs use enameled copper wire around a glass capillary tube or microscope slide.[164-168] A 1 mm diameter hypodermic needle wound with insulated enameled copper 0.127 mm in diameter has also been employed. The copper serves as a heating coil. A copper constantan thermocouple is inserted in the needle's midpoint and a second thermocouple is embedded in the tip of another needle to act as a reference junction. A distance of 200 mm is maintained between the two needles whose emf is measured on a Pye galvanometer. Another direct contact device consists of a mercury thermometer with copper wire wrapped around half of its bulb.[169,170] The wire is heated, and the time required to attain a uniform temperature increase is assumed to be proportional to the soil moisture.

A variety of modified direct contact probes have been developed.[170-173] The probe (Figure 1.22) usually consists of a heater wire with a low temperature resistance coefficient and either a thermocouple or thermistor for temperature measurement. The heating wire is often enclosed in a protective sheath with high thermal conductivity; a thermocouple may be mounted externally. A number of design criteria regarding the probe include a length to diameter ratio greater than 25, guaranteed uniform heating along the probe length, the use of a small diameter heater and thermocouple wire to reduce heat leaks, and the placement of the thermocouple or thermistor as close as possible to the heating element. Potentiometers and galvanometers are most often used for measurement.

In addition to these three electrothermal sensor types, several direct and modified contact probes have been developed for use in open boreholes.[174-176] One such probe is placed in an open borehole or trench side wall (Figure 1.23). The tapered stainless steel point contains the thermistor and heating element which is pushed into the soil.

The theoretical considerations developed primarily for electrothermal probes are well established.[178-187] Assuming an infinite length for a line source of heat, the temperature increase (T) at a radial distance (r) from the source within an isotropic medium with a thermal conductivity (k) is:

$$T = \left(\frac{q}{4\pi k}\right)\left\{-E_i\left(\frac{-r^2}{4at}\right)\right\} \tag{1.9}$$

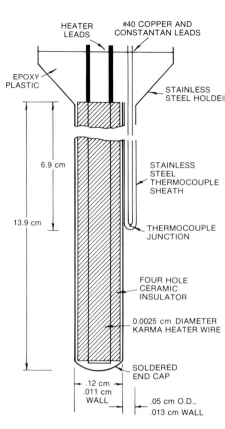

Figure 1.22 A modified direct contact probe.[172]

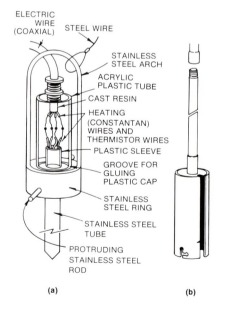

(a)　　　　　　　(b)

Figure 1.23 (a) Modified direct contact electrothermal probe and (b) insertion recovery apparatus.[177] (reproduced from SOIL SCIENCE SOCIETY OF AMERICA JOURNAL, Volume 43, 1979, page 1056, by permission of the Soil Science Society of America)

where   q = the amount of heat produced per unit time per unit length of source (J mm^{-1}),

k = thermal conductivity of the matrix surrounding the source (W m^{-1} K^{-1}),

a = thermal diffusivity (m^2 s^{-1}), and

t = time (s), where

$$-E_i(-K) = \int_K^\infty \frac{1}{K} \exp(-K)\, dK \qquad \text{(1.10)}$$

and K= r^2/4at. In expanded form, this is   **(1.11)**

$$-E_i(-K) = -0.5722 - \ln K + \frac{K}{1 \cdot 1!} - \frac{K^2}{2 \cdot 2!} + \cdots$$

which, for small values of r^2/4at corresponding to small radii or long times, becomes;

$$T = \frac{q}{4\pi k}(b + \ln t) \qquad \text{(1.12)}$$

where (b) is a time-independant constant.

When the source of heating results from an electrical current, then

$$q = i^2 R \qquad \text{(1.13)}$$

where   i = current (A) flowing through the source, and

R = the resistance per unit length of the line source (ohm m^{-1}).

As indicated by equation 1.9, both (k) and (a) will be affected when the porosity around the heating element changes. For this reason, materials which maintain their porosity (e.g., ceramic and Castone) during wetting and drying cycles are preferred. Close contact between the probe and the soil is important to maintain a consistent pore space around the unit.

Installation of in situ porous block and direct contact sensors requires that a hole is augered to the desired depth and the sensor placed firmly within the indigenous soil or side wall. The hole is backfilled with soil and sealed with bentonite at the surface. Contact between the sensor and soil is critical. When air space between the unit and ambient soil occurs, a heat insulating effect develops which minimizes heat transfer and results in an abnormally high temperature increase. This phenomenon is especially pronounced in dry soils.

Once the unit is installed, time is required for the establishment of equilibrium conditions. Resistance (ohms) as a function of time is recorded over a period of 10 to 30 minutes. Upon collection of sufficient data, the heater is turned off and the temperature time relationship is recorded during the cooling cycle. A minimum of 1 hour is recommended prior to a second measurement to allow the soil to return to its original temperature equilibrium. Local drying of the soil (especially if moist) will also occur between successive readings when a sufficient time interval has not elapsed.

Modified direct contact probes are installed permanently or may be withdrawn and repositioned. Since most modified designs are probe shaped, their burial should be horizontal (e.g., an isothermal position) to reduce any temperature variations along the probe. In compressible soils, a hollow tube is used rather than a solid rod to create a hole to withdraw the soil core. Measurements are taken and the probe is retrieved. The sensor in Figure 1.23 allows a hole to be augered above where readings are to be obtained. The stainless steel tube is pushed into the unconsolidated soil thereby minimizing the creation of an annulus between the probe and soil. A steel rod with a diameter slightly smaller than the probe is used to create the horizontal hole in which the sensor is inserted. A special locking device (Figure 1.23b) is connected to threaded rods to position the probe at depth. A wire connected to the sensor or the locking device is used for probe withdrawal.

Electrothermal methods provide a rapid means of soil moisture determination for a wide range of moisture conditions. The approach is independent of salinity effects up to 100,000 ppm for either saturated or unsaturated soils. Although individual sensor sensitivities may vary,[188] reproducible results to within ±5% in the field can be expected.

Establishment and maintenance of thermal contact between the soil and probe is a major limitation with electrothermal techniques.[189] In high shrink swell soils, intimate contact between the soil and sensor is lost with decreasing moisture content resulting in erratic readings until the shrinkage limit is reached. Consistent results, however, have been achieved in sandy soils. The changing heat capacity of the soil water system can also produce small measurement errors. The transfer of heat from the heating element results in moisture movement from the probe although this may be reduced by careful regulation of heat application and heating duration.

c. Nuclear Methods

The three nuclear approaches for soil moisture determination are neutron thermalization, gamma ray attenuation, and nuclear magnetic resonance.

(i) Neutron thermalization

Neutron thermalization methods depend upon the ability of hydrogen in water to slow fast neutrons. The hydrogen count is then correlated with the water content for soil moisture determination.

Neutron thermalization equipment consists of a source of fast neutrons (e.g., energies greater than or equal to 1 million electron volts), a shield, a detector of slow neutrons (e.g., energies between 0 and 1000 electron volts), a scaler or ratemeter, and access tubing (Figure 1.24). A source of fast neutrons can be obtained by mixing an emitter of alpha particles with beryllium. Americium beryllium (Am-Be) and radium beryllium (Ra-Be) are common sources,[191,192] though polonium beryllium and plutonium beryllium have been used[193] The strength of the radioactive source varies; Am-Be in concentrations of 10, 30, 50, and 300 millicurie (mc) have been reported, [194-197] while Ra-Be strengths of 1, 2, and 8 mc have also been used.[198,199] Source materials with long half lives are preferred as the source strength will not significantly fluctuate due to radioactive decay. Fast neutrons emitted from the source create an ellipsoid cloud of neutrons about 7.6 to 15 cm in diameter for wet soils and about 25.4 cm or more for dry soils.

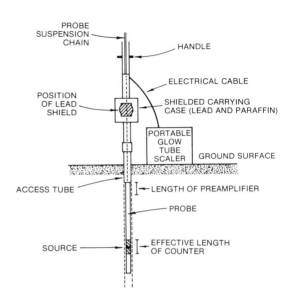

Figure 1.24 Neutron thermalization equipment.[190] (reproduced from SOIL SCIENCE SOCIETY OF AMERICA PROCEEDINGS, Volume 24, 1960, page 436, by permission of the Soil Science Society of America)

The shield provides storage and protection from radiation emitted from the probe which houses the radioactive source and associated recording equipment.[200,201] A shield either includes a standard absorber for equipment calibration or provides a standard count when the data are recorded on the basis of a standard count ratio. Lead and a hydrogenous substance such as paraffin or polyethylene are often used as shielding materials.

A slow neutron detector measures the fast neutrons which have been slowed (thermalized) through elastic collisions with low atomic weight atoms in the soil. Neutron detection equipment consists of a preamplifier, scaler, and timing unit. The most common design has a boron trifluoride (BF_3) proportional counting tube enriched with metallic boron (^{10}B) incorporated into the source probe under pressure. As the thermalized neutrons are absorbed by ^{10}B, an alpha particle is emitted. The alpha particle moves toward a charged wire within the probe, resulting in an electrical pulse which is amplified by a preamplifier in the probe. The number of pulses can be transmitted and recorded with a conventional scaler for a prescribed time interval or with a count ratemeter. A scaler usually permits greater precision while a ratemeter provides a more rapid means of measurement (some models include both). Most scalers include an amplifier, discriminator, and timer.

A variety of probe shapes are available[202-206] whose operational characteristics are highly dependent upon probe geometry. Two basic designs are a surface and a depth probe. The latter is lowered to the desired depth through an access tube. The surface probe (Figure 1.25) can provide measurements to a depth of 40.6 cm in dry soil and 12.7 cm in wet soil.[208-211] The typical surface probe design consists of a cast iron reflector blanketing the source and surrounding the detector. A paraffin or polyethylene shield covers the source and detector to partially compensate for the soil air discontinuity.

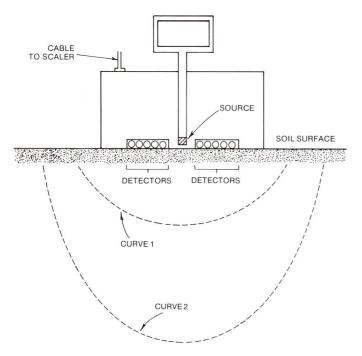

Figure 1.25 Cross-section of a surface probe and subsurface radius of influence; the volumes measured correspond to a moisture content of 43.1 percent of CURVE 1 and 50 percent of CURVE 2.[207] (reproduced from CANADIAN JOURNAL OF SOIL SCIENCE, Volume 41, 1961, by permission of the Agricultural Institute of Canada, Ottowa, Canada)

For a depth probe, access tubes are required to maintain hole stability and insure reproducible hole geometry. Aluminum is the most common casing material[212,213] and is relatively unaffected by neutron flux. Temperature effects upon soil moisture measurements appear to be minimal with aluminum.[214] PVC casing lowers the counting rate since the chlorine content in the polyethylene tubing enhances the counting rate.[215] When installing the access tube, a slightly undersized hole is drilled and the access tube is forced into the hole to ensure intimate tube and soil contact. A small collar placed about 15 cm deep around the access tube eliminates surface water channeling along the tube. For multiple deep well readings, a motorized pulley system is used to lower the probe at a constant rate.

The major factors affecting fast neutron thermalization are the transfer of energy associated with each collision and the statistical probability of the collision. Energy transfer between the fast neutrons and the other nuclei may be represented by:

$$V_c = \frac{M}{m + M} V_0 \qquad \text{(1.14)}$$

where   m = neutron mass,

M = collision nucleus mass,

V_c = energy (velocity) of the neutron after collision, and

V_0 = energy of the neutron before collision.

The number of collisions required to slow a fast neutron can be calculated[216] for any element. The statistical probability of collision is dependent upon the concentration of the element in the soil and its nuclear cross section; an increase in either factor will raise the statistical probability for collision. Soil elements with high nuclear cross sections include lithium, chlorine, boron, and cadmium. When these elements exist in high concentrations in the soil, thermalization can be significantly affected.

While calibration curves are usually supplied by the manufacturer, individual probe calibration is recommended. Calibration curves may be developed from laboratory data, field data, or from theoretical considerations. In laboratory calibrations, large containers of oven dried soil of a known density are mixed with a fixed water volume.[217] Field calibration is usually determined through gravimetric verification.[218-221] Inaccuracies in either testing method may result from the presence of elements with high nuclear cross sections. The presence of hydrogen in organic matter, structural hydrogen in clay, or tightly bound soil hydrogen nuclei which react differently than water hydrogen nuclei can affect the calibration curve of the instrument. Other factors affecting both laboratory and field calibration measurements are soil density, temperature, variations in hydrogenous material other than water, electrical contact resistance, abrupt changes in salt concentration and flux, and swelling and shrinking soils.[222-224] Bias from soil cracks or loose backfill around the access tube can result in significant negative bias of up to 30%; moisture within the access tube can also introduce error. If different access tube materials or diameters are used, calibration is required for each type.[225] A theoretical calibration curve assumes that soil constituents other than soil moisture are negligible in the slowing of fast neutrons.[226]

Field application techniques for neutron thermalization equipment vary according to whether a surface or depth probe is employed. Measurements with a surface probe require the probe to be placed firmly in contact with the soil by scraping or pressing it against the soil. The soil surface is carefully smoothed prior to placement of the probe as failure to obtain complete contact will result in unreproducible data containing significant error. Pronounced soil stratification can also introduce error.

Access tubes are needed for equipment utilizing a depth probe. Shallow access tubes may be installed with augering followed by insertion of the access tube. This method ensures a straight hole and minimizes air pockets between the access tube and soil. Deeper access holes drilled with conventional drilling techniques present such problems as water introduction to the hole (e.g., drilling muds), development of air cavities around the access tube, and inadequate backfilling procedures.[227] The access tube should be plugged prior to insertion if there is any possibility of water entering the tube. Upon completion of the access hole and equipment standarization, the probe is lowered into the well with measurements commencing at a minimum of 19 to 25 cm from the surface. Measurements closer to the surface can result in the escape of fast neutrons into the air unless a reflecting shield is used.

Neutron thermalization provides a rapid method of obtaining soil readings which are largely independent of temperature and pressure. An average moisture value for water in the solid, liquid, or vapor state is quickly determined. Replicate measurements can be easily recorded and integrated into an automated data acquisition system. Changes in the soil moisture content can also be observed immediately. Disadvantages include measurement bias from certain elements in the soil and the difficulty in defining the horizontal distribution of water near the probe, since moisture close to the neutron source has a more pronounced effect on the counting rate than pore water at a greater distance.[228]

(ii) Gamma ray attenuation

The attenuation of a nearly parallel beam of gamma ray radiation through the soil can be used to determine soil moisture content. Since the absorption of gamma rays is primarily a soil density dependent phenomenon, the method is not water specific but relies on the change in effective density attributable to changing moisture content.

Three basic equipment designs are available which are categorized according to the type of gamma source probe. These are a double probe (transmission method), and a single probe configuration (scattering method). The scattering method is differentiated as either a single or double radiation source.

The double probe design is illustrated in Figure 1.26. One probe contains the gamma photon source that is usually ^{137}Cesium (Cs), ^{60}Cobalt (Co), or ^{241}Americium (Am). For most purposes, 20 or 25 mc of gamma radiation is satisfactory though 100 to 500 mc may be used for higher resolution.[230,231] The gamma ray source is housed in a lead or tungsten shield with a collimating hole or slit in the shield casing.[232,233]

The second double probe design contains a detector, photomultiplier, and preamplifier.[234] Detectors generally consist of lithium germanium Ge(Li) or thallium activated sodium iodide NaI(Tl) scintillation crystals which are connected to a photomultiplier tube and preamplifier. The Ge(Li) detector provides a higher attainable resolution than the NaI(Tl) detector although a variety of scintillation crystals have been used.[235] Output from the scintillation counter (e.g., scintillator confined with a photomultiplier tube) is often connected to a linear amplifier and then to a scaling circuit. A pulse height analyzer (e.g., discriminator) sorts pulses from the detector according to size, and records the pulse on the appropriate channel. This process screens secondary radiation

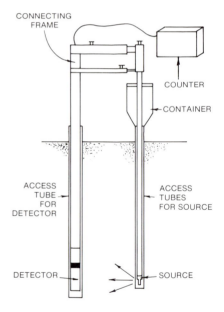

Figure 1.26 Double probe gamma transmission.[229] (reproduced from JOURNAL OF HYDROLOGY, Volume 9, 1969, by permission of the Elsevier Scientific Publishing Co., Amsterdam, Netherlands)

resulting from the interaction of the primary gamma rays in the soil.

The single probe design contains the gamma source and detector separated by a polyethylene or lead shield (Figure 1.27).[237] Radiation from the source passes into the soil and a portion is scattered back to the detector.[238] A modification of the single probe design contains two gamma sources in one collimator and a detector in the second collimator. Bulk density and moisture content can be measured concurrently when ^{137}Cs is used as a high energy source and ^{241}Am as a low energy source.[239-244]

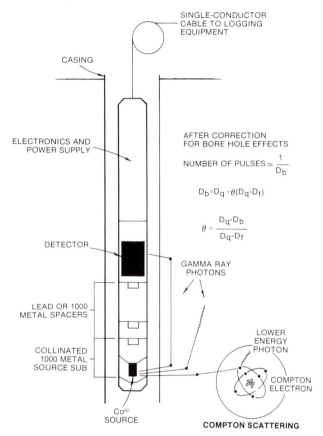

Figure 1.27 Principles and interpretation of single probe gamma transmission equipment.[236]

The attenuation or scattering of monoenergetic gamma rays for a fixed source detector arrangement is affected by three major phenomena. These are: (1) Compton scattering, in which partial photon energy is lost to an orbital electron; (2) photoelectric effects, in which an ejected orbital electron completely absorbs the photon energy; and (3) pair production, which occurs as the photon approaches the nucleus and completely converts itself into a pair of electrons. Pair production and photoelectric effects are negligible for energies slightly over 1 MeV. Compton scattering or absorption is proportional to A/Z where A is the atomic weight and Z is the atomic number. For all elements except hydrogen, Z is constant until Z equals 30 on the periodic table. Since the range of Z from the atomic numbers 2 to 30 includes most common soil elements, absorption is independent of soil chemical composition for these energies.[245]

Attenuation of gamma rays depends upon the total soil density.[246,247] If the value of soil density minus its water content is constant, then the resulting attenuation reflects changes in the soil moisture content. The basic equation describing this attenuation is:[248,249]

$$\frac{N_m}{N_o} = \exp\ [\ -S(\ \mu_s\ \rho_s\ +\ \mu_w\ \theta\ -\ 2S'\ \mu_c\ \rho_c)\ ] \tag{1.15}$$

where

N_m/N_o = ratio of the transmitted gamma ray to the incident flux for moist soil,

S = thickness of soil column (cm),

μ_s = mass attenuation coefficient for the soil (cm²/g),

ρ_s = soil bulk density (g/cm³),

μ_w = mass attenuation coefficient of water (cm²/g),

θ = water content per unit bulk volume of soil (g/cm³),

S' = thickness of container walls (cm), and

μ_c = mass attenuation coefficient for container material (cm²/g).

The corresponding dry soil equation is:

$$\frac{N_d}{N_o} = \exp\ (-S\ \mu_s\ \rho_s\ -2S'\ \mu_c\ \rho_c) \tag{1.16}$$

Equation (1.15) divided by (1.16) becomes

$$\frac{N_m}{N_d} = \exp\ (-\mu_w\ \theta\ S) \tag{1.17}$$

which yields $\theta = \dfrac{\ln(N_m/N_d)}{-\mu_w S} = \dfrac{\log\ (N_m/N_d)}{-0.4343\ \mu_w S} \tag{1.18}$

Equation (1.16) assumes that attenuation due to air within the soil column is negligible due to the low density of air within the soil column. Absorption by the casing material (S') is similarly considered to be minimal. The equation also requires that for water content determinations, the soil density must be known for a fixed water content value. Changes due to compaction with depth (e.g., an apparent increase in water content) or soils exhibiting a high shrink swell factor are assumed to be negligible.

Prior to obtaining the water content value (θ), it is necessary to evaluate the terms for the mass absorption coefficient for water (μ_w) and for the count rate through a dry soil (N_d). Values for μ_w may be obtained either from gamma ray absorption coefficient tables[250] or empirically with a representative soil. N_d may be obtained on either an oven or air dry basis.

Field installation and measurement with the double probe unit requires installation of two parallel access tubes. The detector and source probe are lowered concurrently into the two access tubes; a guide is used to ensure that the units remain parallel. Collection of accurate soil moisture data depends more upon the tube and probe being parallel than any other factor. Drill stem and guide equipment for access tube installation, positioning assemblage for both probes, and mechanical depth controllers are described in the literature.[251-253]

Field measurements with the single probe design are obtained by lowering the unit down an access hole. Because the length of the single probe is separated by several centimeters of lead, the smallest vertical thickness of soil which can be surveyed is governed by this distance. Readings should commence at least 15 cm from the surface with this design to minimize error from surface radiation loss.

Gamma ray absorption provides accurate water content measurements that are unaffected by time lag. High precision with relatively small gamma ray sources are possible for a layer smaller than is possible with neutron thermalization. Field tests have shown that an effective measuring width of 1 to 2 cm is obtainable. A dual gamma source in a single probe can provide both soil density and water content concurrently, and is therefore unaffected by swelling soils and freezing and thawing cycles. The reading obtained with any of the designs provides a measurement of the water content irrespective of its physical state.

Limitations include the cumbersome nature of the equipment, difficulty of proper alignment of the access tube(s), and the increasing possibility of inaccurate soil density values over depth and time. While the approach can precisely determine moisture content changes, the absolute moisture content may be subject to large errors. Measurements obtained near a wetting front can be inaccurate because the wetting front represents a portion of the soil profile that includes both wet and dry soil. Major errors also occur with depth, presumably due to increasing soil density; anisotropic soils with swelling and shrinking characteristics similarly introduce bias. Most equipment problems associated with temperature, Compton scattering, discrimination of scattered and secondary radiation, count rate instability, and electronic drift can be solved with additional equipment, calibration, longer counting times, or more intense gamma radiation sources.

(iii) Nuclear magnetic resonance

Nuclear magnetic resonance (NMR) is based upon the interaction between nuclear magnetic dipole moments and a magnetic field. Transitions in the energy levels of the atomic nuclei created by the oscillating electromagnetic field of an appropriate frequency result in the energy absorption from the electromagnetic field by the nuclear magnetic dipole system. The electronic detection of such energy absorption or corresponding nuclear dipole excitation results in a nuclear magnetic resonance signal. Detection of the NMR signal produced by the hydrogen nuclei of a water molecule therefore provides a direct measurement of the hydrogen nuclei concentration in the soil sample.

While several methods for producing and detecting NMR signals are available, the transient or pulsed method is preferred for soil moisture measurements since readings can be conducted rapidly and the resulting data provides extensive information on the nuclear species of interest.[248, 254-262] In this method, a pulse is applied in short bursts with the magnitude and duration determining the angle with respect to the static magnetic field that the nuclear magnetization vector will be reorientated. In general, the angle through which the magnetization is rotated is:

$$\theta = \gamma H_1 \tau_w \qquad \text{(1.19)}$$

where H_1 = amplitude of the rotating radio frequency magnetic field, and

   τ_w = pulse duration.

The angle (θ) most commonly used is either 90° or 180°.

A variety of NMR designs have been proposed, although the basic components are similar. These are a source for creating an applied static magnetic field (a magnet), a radiofrequency induction coil, and the associated electronic circuitry which is connected (a coaxial cable) to the radiofrequency coil. A schematic of one sensor is shown in Figure 1.28. The soil sample (1) is surrounded by a radiofrequency induction coil (2) in which a magnetic field (3) of strength (Ho) is applied. A radio frequency oscillator (4) supplies pulses of energy to the coil for the proper duration and separation to reorient the spin axis of the hydrogen nucleus. The voltage is increased by the amplifier (5), demodulated by the detector (6), and displayed as a function of time on a cathode ray oscilloscope (7). A sampling gate (8) and digital voltmeter (9) provide a readout of the amplitude of the nucleus reorientation.

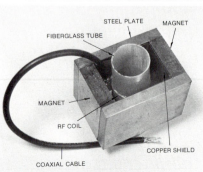

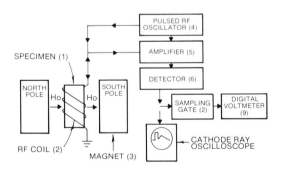

Figure 1.28 Components of a pulsed nuclear magnetic resonance sensor and associated instrumentation.[262]

Figure 1.29 Nuclear magnetic resonance sensor.[148]

A field prototype (Figure 1.29) contains a radiofrequency transmission and detection coil wound around a plastic tube. The fiberglass tube is placed within a copper shield which is cast into an acid resistant plastic case. The sensor (8.9 x 6 x 7 cm) which weighs 1.2 kg, is connected to a coaxial cable for signal transmission.

The theoretical basis for NMR is described in the literature.[263-265] The measured NMR signal amplitude is proportional to the mass of moisture within the effective volume of the detection coil by:

$$\xi = Cm_v V \qquad \text{(1.20)}$$

where $\quad \xi$ = signal amplitude (volts),

$\qquad$ C = a proportionality factor (also called the gyromagnetic ratio) between the measured NMR signal and the water mass,

$\qquad m_v$ = moisture density (g/cm³), and

$\qquad$ V = volume of the coil (cm³).

The proportionality constant (C) is dependent upon a number of factors including the geometry of the detection coil, the filling factor, the gain of the detection electronics, the strength of the applied magnetic field, and the magnitude of the nuclear magnetic moments and temperature. The NMR signal amplitude is therefore related to the dry weight moisture fraction by:

$$\frac{\xi}{D} = K \frac{m}{1 + m} \qquad \text{(1.21)}$$

where $\quad$ K = a proportionality factor,
$\qquad$ D = density of the soil plus moisture content (g/cm³), and
$\qquad$ m = dry weight moisture fraction.

For soils with various moisture contents, volumes, and densities, ξ/D is expected to be linearly related to the quantity $m/(1+m)$. For situations where the water molecule is absorbed on the surface of a soil particle or is chemically bonded to the solid constituents, the equations can be modified.

Calibration curves are needed for each soil type of interest. The moisture content of a representative soil is determined gravimetrically followed by measurement with an NMR sensor. Depending upon the geometry of the detection cavity, sealed containers with the wetted soil can be placed directly into the sensor. The NMR values versus the volumetric moisture content (Equation 1.20) are then plotted.

A temperature correction curve may also be required since NMR readings vary inversely with temperature in a linear fashion. Since temperature affects the gain of the electronic readout system, the entire unit can be placed within an oven and the temperature adjusted over the range encountered in the field. Readings obtained from a sensor placed within the oven can be used to develop this curve.

Installation procedures for NMR sensors depend primarily upon the geometry of the unit. A technique developed for the Figure 1.29 sensor involves driving a thin walled plastic tube into the bottom of a test pit.[148] Soil is then removed from around the tube to a depth of about 5 cm. The sensor is slipped over the tube and seated firmly in the test pit. The excavation is backfilled and compacted to a density similar to the surrounding soil. Plastic sheeting is used to line the test pit (excluding the area for sensor installation) and the excavated soil to reduce moisture loss during installation.

NMR techniques offer a rapid means for determining the moisture content in the 0 to 50% range. The method is independent of bias from salt concentrations up to 2000 ppm, organic matter, and soil temperature (assuming temperature correction). An estimated accuracy of ±2% is available with current sensors. Disadvantages include the dependence of the device on differences on the bonding mechanism of water in different soils. Soils containing non-water hydrogen, variations in measurements between bound and free water, and signal overlap from hydrogen containing liquids such as fats or oils pose additional problems. Further equipment refinement and field research is required to determine the accuracy and reliability of the method before its use becomes widespread.

C. SOIL SALINITY

The soluble salt concentration of a soil can be a valuable pollution indicator. Most laboratory and field techniques which measure soil salinity rely upon the relationship between electrical conductivity or resistance and salt content. Techniques using this principle include the laboratory extraction of a soil sample and measurement of the pore water conductivity, collection of soil pore water and subsequent analysis, salinity sensors, the four electrode method, electrical conductivity probes, and electromagnetic approaches. Appropriate laboratory procedures are described in the literature[266,269] while pore water sampling is addressed in Section E of Part I. The remaining approaches are presented herein.

1. Salinity Sensors

A salinity sensor provides an indirect method of expressing soil salinity in terms of electrical conductance. When a porous insulator buried in the soil attains diffusion equilibrium with the surrounding soil pore water, the salt concentration within the insulator is presumed to be representative of the surrounding soil. A conductivity measurement of the pore water in the insulator is therefore used to determine the soil salinity in the immediate vicinity of the sensor.

A variety of sensors have been developed. An early version consisted of two platinum wires embedded in porous ceramic.[270] An improvement to this design has platinum screen mesh electrodes of dissimilar surface areas fixed on the opposite sides of two parallel 1 mm thick ceramic plates.[271] A thermistor is included in the unit so that electrical conductivity measurements are referenced to a standard temperature. To ensure intimate soil contact with the electrolytic element, a spring mount is incorporated into the sensor. The spring is activated by withdrawing a pullwire from the unit which forces the electrolytic element against the soil. This feature is highly desirable in soils of questionable mechanical stability; an identical unit without the spring can be used in more cohesive soils. A cross section of the spring mounted version is illustrated in Figure 1.30. A sensor based upon this design uses ceramic as a casting agent. This construction prevents electrode dislodgement and improves sensor stability.[273] Temperature coefficients obtained from published tables[266,274-276] are used for temperature compensation as a thermistor is not built into the unit.

Other sensor shapes include probes, cylinders, and squares. Figure 1.31 shows a probe design made with a plexiglass core and an inner and outer platinum wire electrode. The cylindrical and square units in Figure 1.32 contain platinum electrodes cast in a porous glass. A thermistor is included in the sensor and compensates for soil temperature effects between 5 and 35°C.[278]

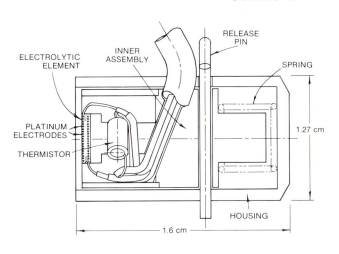

Figure 1.30 Spring loaded salinity sensor.[272] (reproduced from 5000-A SOIL SALINITY SENSOR, 1980, by permission of Soilmoisture Equipment Corp.)

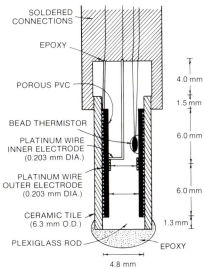

Figure 1.31 Plexiglass soil salinity sensor.[277] (reproduced from SOIL SCIENCE SOCIETY OF AMERICA PROCEEDINGS, Volume 34, 1970, page 215, by permission of the Soil Science Society of America)

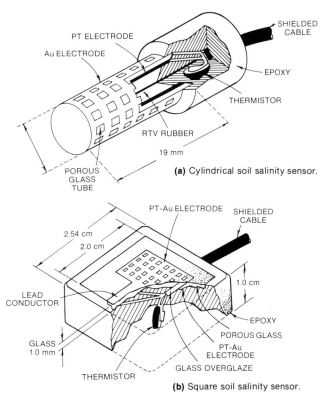

(a) Cylindrical soil salinity sensor.

(b) Square soil salinity sensor.

Figure 1.32 Cylindrical and square soil salinity sensors.[278] (reproduced from SOIL SCIENCE SOCIETY OF AMERICA PROCEEDINGS, Volume 33, 1969, page 787, by permission of the Soil Science Society of America)

Salinity sensor measurements are usually recorded with an a.c. Wheatstone bridge.[279] Other salinity bridges incorporate a galvanometer null indicator and an electrical analog to provide temperature corrected (to 25°C) conductivity measurements. A computerized system of stepping switches can be used to sequentially obtain measurements from a number of sensors with temperature correction.[280] Another approach uses a variable frequency oscillator circuit to linearly alter the frequency with electrical conductance.[281]

In general, a linear relation exists between sensor conductance and the conductivity of the calibration standard (KCl solution at 25°C). The exception is a curvilinear response for salinities below approximately 1.5 mmhos/cm.[282] If sensor conductance (Cs) equals 1/Rs (resistance), the equation of this line is:

$$C_S = S(\gamma - A) \tag{1.22}$$

where S = slope at 25°C,

   A = conductivity intercept, and

   γ = conductivity value for a KCl standard
       solution at 25°C.

If a negative A value occurs, the a.c. resistance (Rs) becomes:

$$(\gamma - A) = \frac{1}{S \cdot Rs} \tag{1.23}$$

where $A = 0$ for conductivity cells in bulk
       solution, and

   $1/S$ = cell constant.

Since the conductivity values (γ) change approximately 2% per °C, a thermistor or a temperature correction curve can be developed directly from the resistance (Rs).

The sensor is calibrated by placing it in a standard salt solution of a known electrical conductivity at 25°C. A minimum of 24 hours is recommended prior to measuring the conductance to allow the sensor to reach equilibrium with the bulk solution. A graph of sensor conductance versus solution conductivity is then constructed. Conductance measurements should be performed with the same Wheatstone bridge as differences between bridges can vary as much as 4%. Sensor recalibration is recommended every two to three years if precise salinity measurements are required.[283]

Installation procedures require that soil removed from the borehole is placed sequentially at the ground surface and not mixed. Once the sensor is placed at the proper depth, the corresponding backfill is replaced and compacted every 10 to 20 cm. Because soil disturbance above the sensor affects soil permeability, maintaining a backfill permeability either equal to or greater than that of the undisturbed soil is desirable. A mixture of finely powdered bentonite and excavated soil is used in the latter case. Electrolyte equilibrium between the sensor and soil solution requires a minimum of 24 hours prior to measurement although mathematical calculations, based upon the response factor of an individual sensor, may be used to correct these readings to reduce this initial and subsequent time lag response.[284]

Salinity sensors measure soil salinity without continuous soil disturbance and are conducive to automated recording systems. Most sensor readings may deviate from actual conditions by ±0.5 mmho/cm. Disadvantages include the inability to precisely monitor soil salinity in dry soils with a moisture tension of approximately 2 bars,[285] declining sensor accuracy at low soil matric potentials, temperature dependence of the method, and the slow response time associated with rapidly changing salinities.

2. Four Electrode Method

The four electrode method relies upon the conductance or resistance measured between a known current, which is passed through the primary circuits of two electrodes, and a secondary circuit, which connects two inner electrodes. The four electrodes are usually placed on the ground surface[286-288] although units have been buried for determination of bulk soil salinity.[289] An idealized four electrode surface installation is depicted in Figure 1.17; the basic approach and equipment are identical to the four electrode method used for determining soil moisture content.[290] Probes are often constructed from brass, stainless steel, copper, or aluminum. Portable meters are available for direct measurement of conductivity or resistance with temperature compensation.[291-294]

Theoretical considerations for the four electrode method are well established and the detailed governing principles for this method are available in the literature.[295-297] For a soil medium, the resistivity (ρ) is defined as:[298]

$$\rho = \frac{4\pi aR}{n} \tag{1.24}$$

where   a = the electrode spacing,

R = measured resistance, and

n = a boundary condition (e.g., soil surface).

For an evenly spaced electrode array, the boundary condition (n) is given as:

$$n = 1 + \frac{2}{\sqrt{(1 + 4\lambda^2)}} - \frac{1}{\sqrt{(1 + \lambda^2)}} \tag{1.25}$$

where $\lambda = b/a$, and

b = depth of electrode beneath the soil surface.

For large (i.e., in comparison to a) values of b,

$$\rho = 4\pi aR. \tag{1.26}$$

When b is small in relation to a, resistivity (ρ) becomes

$$\rho = 2\pi aR. \tag{1.27}$$

From these general considerations, the four electrode method was developed to evaluate the relationship between conductivity and soil salinity at different moisture contents.[299] Soil electrical conductivity (EC_W), as measured by conductivity methods, depends upon the electrical conductivity of the soil pore water (EC_w), the volumetric water content (θ), soil pore geometry or tortuosity (T), and surface conductance (EC_S).[300] This relation is given by:

$$EC_a = (EC_w \, \theta) \; [T] \; + EC_s \qquad \text{(1.28)}$$

where (T) is an empirically developed, dimensionless transmission coefficient dependent upon the volumetric water content (θ) and given by:

$$[T] = a \, \theta + b \qquad \text{(1.29)}$$

where   a,b = linear regression constants.

Since the soil type influences (T) and (EC_S), the apparent soil electrical conductivity (EC_a) for any given soil type becomes:

$$EC_a = A_1(EC_w \, \theta) + B \qquad \text{(1.30)}$$

where A_1 = T, and
$B = EC_S$.

If EC_a measurements are obtained for a fixed water content,

$$EC_a = A_2 \, EC_w + B \qquad \text{(1.31)}$$

and $$EC_a = A_3 \, EC_e + B \qquad \text{(1.32)}$$

Calibrations for the four electrode method are obtained using field methods, a four electrode cell, or an electrical conductivity probe.[301,302] In the field approach, numerous readings at an interelectrode spacing of 30.5 cm are taken. Representative soil samples are collected and their salinity obtained via saturated paste extracts. A calibration curve is then constructed. In a four electrode cell (Figure 1.33), a representative soil core is placed within the cylinder and measurements are obtained with electrodes attached to the cell wall. A standard electrical conductivity solution is used to fill the cell to obtain the cell constant (K) and to derive the cell resistance (R_T) by:

$$K = EC_{25} \cdot \frac{1}{f_T} \qquad \text{(1.33)}$$

where f_T = a temperature correction factor (to 25°C).

The sample is removed and the soil salinity of the saturated paste extract is measured. The conductivity of this paste versus the apparent electrical conductivity is then graphed. Calibration procedures with a conductivity probe are performed by calibrating the probe followed by field measurements. Four electrode readings are then correlated with the data obtained from the conductivity probe.

The four electrode method involves placing the four electrodes in the soil surface along a straight line to maintain a consistent distance between the electrodes. An inaccurate spacing, especially between the middle electrodes, can result in significant error. Surface electrodes are therefore often arranged in a fixed jig; the burial of in situ electrodes similarly requires a guide to ensure proper vertical and horizontal placement. The interelectrode distance (the "a" spacing in Figure 1.17) approximately equals the depth of the

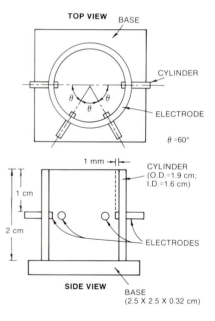

Figure 1.33 Four electrode conductivity cell.[303] (reproduced from SOIL SCIENCE SOCIETY OF AMERICA JOURNAL, Volume 40, 1976, page 152, by permission of the Soil Science Society of America)

resistivity measurement. Varying the "a" spacing thus yields the bulk soil conductivity for different soil volumes. The apparent soil conductivity (EC_w) in mmho/cm is:

$$EC_a = \frac{5.222}{a} \cdot \frac{f_T}{R_T} \qquad (1.34)$$

where R_T = resistance in ohms for an
equidistant interelectrode spacing,

f_T = temperature correction factor to 25°C, and

a = electrode spacing.

The four electrode method provides an average bulk salinity value for a large area and excellent soil salinity and electrical conductivity correlation. The approach also eliminates soil disturbance by taking replicate measurements at fixed time intervals. Permanent installation of four electrodes in the soil allows determination of salinity variations over time. However, installation results in disturbance of the soil column, and the precise placement of several sets of electrodes at a fixed "a" spacing is critical. Differences in soil texture, temperature and fluctuating soil moisture content will also affect the measurements.

3. Electrical Conductivity Probes

An electrical conductivity probe measures the conductivity of the soil pore water. The two types of probes used for this purpose are classified as either portable or burial.

A portable or burial probe consists of four annular rings molded into a 12 cm long plastic tube. In the portable probe variety, lead wires from the electrodes are connected through an aluminum shaft handle to a resistivity meter. The probe is inserted into a borehole, readings are taken, and the probe is retrieved. In the burial version, the probe is buried and the electrical wires are led to the surface. One burial variety incorporates neutron scattering equipment in the probe so that the electrical conductivity and moisture content are measured concurrently.[304] The electrodes on either probe type can be spaced to allow measurement of various soil volumes.[305] A current passes through the outer electrodes and the potential current drop between the outer and inner electrodes is recorded.[306,307]

Calibrations of an electrical conductivity probe are obtained by saturating a small volume of soil with a saline solution of known conductivity and measuring the soil conductivity with the probe.[308] Saline solutions (electrical conductivities = 4, 10, 20, and 40; sodium absorption ratio = 8) are prepared for placement into open cylinders 30.5 cm in diameter by 46 cm long, which are driven 10 to 15 cm into the soil. A 15 cm wide moat around each cylinder is excavated, and one of the saline solutions is poured into the cylinder and moat. Once the soil has reached field capacity the cylinder is removed and a 30.5 cm deep hole is augered into which the probe is centered. An apparent electrical conductivity of the bulk soil is obtained and the probe is withdrawn. The electrical conductivity of a paste extract of the soil, collected with a 15 cm barrel auger, is then measured. The electrical conductivities of the soil saturated with the known saline solutions are graphed to establish an apparent relationship between electrical conductivity and soil salinity.

39

A portable probe is installed by augering a hole with a diameter slightly smaller than the probe. The tapered probe is forced into the hole and a soil resistance measurement is obtained (Figure 1.34). Permanent installation of a burial probe also requires augering a hole in which the unit is placed. For installation at depth, lead wires are fed through an electrical conduit tube (Figure 1.35). The access hole is backfilled and compacted to prevent surface water channeling.

Figure **1.34** Resistance measurements from installed portable soil salinity probe.[309] (reproduced from EQUIPMENT FOR SOIL RESEARCH, 1981, by permission of the Eijkelkamp, B.V.)

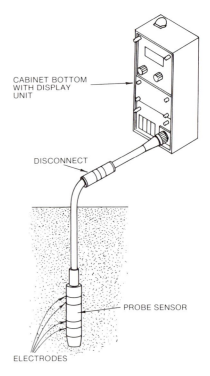

CABINET BOTTOM WITH DISPLAY UNIT

DISCONNECT

PROBE SENSOR

ELECTRODES

Figure **1.35** Installation of in situ soil salinity probe.

An electrical conductivity probe accurately measures soil salinity for a small soil area. Intermittent (portable probe) or continuous in situ (burial probe) monitoring is possible. Disadvantages include the small volume of soil surveyed (limited by the electrode spacing) and soil disturbance from augering.

4. Inductive Electromagnetic Methods

Inductive electromagnetic methods rely upon the creation of a magnetic field which passes through the soil to obtain a bulk electrical conductivity value. Loops of electrical current are used to generate the magnetic field which passes through the soil and is intercepted by a receiver whose measurement is linearly related to the apparent soil conductivity.

Electromagnetic probe designs include surface instruments, held at a prescribed distance from the soil during measurements, and a downhole version. Although probe designs vary in geometry and construction materials, the basic transmitter and receiving components are similar.[310,311] The surface hand held probe design consists of a transmitter coil placed at one end of the instrument (Figure 1.36) which produces a current loop. The intercoil spacing is fixed with the coils in a coplanar

40

fashion. The unit is held parallel to the ground surface at a vertical distance such that a portion of the loop enters the soil. A receiver coil mounted on the frame intercepts a portion of the induced electromagnetic field. The signal from the receiver is amplified and converted to a voltage value which is directly related to the apparent electrical conductivity.

Figure 1.36 Inductive electromagnetic soil conductivity meter. (photograph courtesy of JAMES D. RHOADES, Research Leader, U.S. Salinity Laboratory, Riverside, CA)

The downhole probe consists of two electromagnetic coils and appropriate circuitry and is designed to be inserted into a 5 cm inner diameter access tube. An electrical magnetic field corresponding to 15 cm depth increments is produced. Probe frequencies can be adjusted in three ranges (0-3, 0-10, 0-30 ms/cm) so that the measured soil volume can be varied.

Calibration of the unit is performed by the manufacturer for an apparent bulk soil resistivity value and constant distance between the ground surface and probe. An electrical conductivity probe or soil paste extract is used for additional calibrations which are required for geographical areas other than where the unit was manufactured. Variations in the mineralogy, which appear to affect the magnetic loop of different regions, are compensated with this calibration. Once this is done, surveying can be performed without further calibration.

Surface probe designs require that the unit is held at the calibration height and parallel to the soil surface. The depth of measurement is determined by the distance of the instrument from the soil, transmitter frequency, intercoil spacing, and coil to ground surface orientation. Adjusting the height of the instrument above the soil surface yields the apparent electrical conductivities with variations in depth.[312] Another procedure obtains measurements when the coils are placed parallel and perpendicular to the ground surface. When considered in terms of the appropriate mathematical relationships derived from geophysical instrumentation data, a conductivity depth profile can be established.[313]

Electromagnetic induction is a quick method of surveying a large area at a prescribed depth. While most units are designed for use at relatively shallow depths (<6 m), two men instruments with 3 coil spacings are capable of measurements to 60 m. Additional research is required to refine the method and to overcome problems caused by interference from magnetic materials in the soil and poor depth resolution.

D. TEMPERATURE

Temperature is a basic physical property which can be measured by the response of matter to heat. Any device that has been calibrated to an acceptable standard is considered to be a valid temperature measuring instrument.[314,315] A number of devices can be used to measure temperature in the vadose or saturated zone and include liquid in glass, bimetallic, filled systems, thermocouples, and electrical resistance thermometers.[316] A liquid in glass sensor consists of a thin walled glass bulb connected to a capillary stem which is sealed at the opposite end. The bulb and part of the stem are filled with an expansive liquid such as mercury, mercury thallium, gallium, alcohol, toluol, pentane, or a silicone. The liquid volume expands and contracts with corresponding temperature fluxes.

A bimetallic sensor is composed of two dissimilar metal strips which are rolled to form one metal. Differences between the linear expansion rates of the metal result in changes in metal curvature with temperature. Invar and brass or Invar and steel are commonly used and form a flat spiral, single helix, or multiple helix configuration.

A filled system, also known as a Bourdon thermometer, consists of a bulb connected via a capillary tube to a buried elliptical bulb. A gas or liquid, whose expansion or contraction is transmitted to a Bourdon gage, bellows, or indicating arm, fills the system. The recording apparatus may be sensitive to either volume or pressure changes.

A thermocouple is the most commonly used instrument for soil temperature measurements and consists of two dissimilar metals so joined as to produce a thermal emf when the junctions are at different temperatures. One circuit acts as a measuring junction while the second is a reference junction. A complete sensing assembly consists of one or more of the following: a sensing element assembly (two dissimilar wires), a protecting tube, a thermowell (i.e., a protective tube able to withstand high pressures, stress or erosion), and the termination of the sensing element.[317] Materials, such as copper constantan, platinum, rhodius platinum, and Chromel alumel are used for sensing wires so that a fairly linear temperature emf relationship is produced. A potentiometer, galvanometer, or millivoltmeter measures the thermal emf. Typical thermocouple element assemblies are illustrated in Figure 1.37.

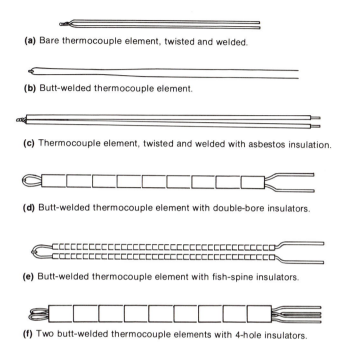

(a) Bare thermocouple element, twisted and welded.

(b) Butt-welded thermocouple element.

(c) Thermocouple element, twisted and welded with asbestos insulation.

(d) Butt-welded thermocouple element with double-bore insulators.

(e) Butt-welded thermocouple element with fish-spine insulators.

(f) Two butt-welded thermocouple elements with 4-hole insulators.

Figure 1.37 Thermocouple element assemblies.[317] (reprinted with permission from the American Society for Testing and Materials, 1916 Race Street, Philadelphia, PA 19103. STP 470B)

Thermocouples wired in a series are called thermopiles or thermels. Thermopiles are constructed so that the measuring and reference junctions consist of several different couples. A data acquisition system records multiple readings and often includes programmable stepping switches for contact closure of the individual units.[318-320]

Electrical resistance thermometers operate on the principle that the electrical resistances of metallic (e.g., a thermistor) and semiconducting materials react to changes in temperature.[321-324] P-N diodes are the most common thermometers, although Nichrome, copper, silver, or platinum wire encapsulated in epoxy are also used.[325-327] An a.c. Wheatstone bridge or potentiometer usually measures the resistance changes due to temperature.

Installation of temperature recording devices vary according to the sensor type. Nevertheless, in all cases, the depth at which the unit is withdrawn or buried must be precisely known. Mercury in glass thermometers are commonly suspended in an access tube at a prescribed depth. To minimize temperature changes during withdrawal, the bulb is encapsulated in a paraffin or micro crystalline wax.[328,329] Bimetallic and filled system thermometers are installed in close contact with the soil, and radiant heating of above surface elements should be avoided to reduce temperature bias. Thermistors and thermocouples should also be firmly installed in the soil to ensure good soil contact. A meter or more of lead wire should be buried to minimize the effects of thermal conductivity due to exposure to different soil temperatures. Similar precautions should be observed for thermopile installation.

The advantages and limitations of temperature measurement depend primarily upon the type of sensor being used. Liquid in glass thermometers exhibit errors attributable to thermal conductivity of the access pipe. Heat convection within the tube can also influence measurements when pronounced soil temperature gradients exist. Bimetallic thermometers are durable and respond rapidly to temperature changes though they are less precise than liquid in glass thermometers. Problems associated with filled system thermometers include their fragile nature and the high sensitivity of the capillary tube to radiant heating.

Thermocouples offer an excellent means of integrating temperature either vertically or horizontally at multiple depths. Unit accuracy depends upon voltage measuring equipment, the amount of electrical noise in the lead wires, and errors due to variations in the reference electrode.

Thermistors are ideal for automatic recording systems, and they provide a high degree of sensitivity and stability, especially when the unit is preconditioned by electrical overheating prior to measurement.[330] Preconditioned units have been found to be sensitive to within 0.001°C for several years after soil installation.[331] To produce this degree of accuracy, the unit must be designed to avoid thermistor self heating due to its bridge circuit.

E. SOIL PORE WATER SAMPLING

Soil pore water samples are obtained either by extraction of a soil sample or with in situ samplers. In the former method, a soil core is collected from which pore water is withdrawn by displacement, compaction, centrifugation, molecular adsorption, or suction. In situ techniques rely upon creating a vacuum to induce soil pore water flow into a collection vessel. Sampling devices developed for this purpose include vacuum pressure lysimeters, vacuum plates and tubes, membrane filter samplers, and absorbent devices.

1. Vacuum Pressure Lysimeters

Vacuum pressure lysimeters collect soil pore water by creating a vacuum within the sampling vessel; pore water moves toward the sampler and enters the vessel through a porous section of the lysimeter. The type of hydrophilic material used in a lysimeter provides a convenient method of categorization and includes four porous materials: ceramic, fritted glass, nylon, and Teflon®.

Irrespective of the particular lysimeter design or selection of hydrophilic material, a number of inherent limitations exist with this method. These problems include the uncertainty of the degree to which the collected sample represents the surrounding pore water, the disruption of normal drainage patterns caused by suction induced sampling, clogging, and the potential sample contamination from materials used in the lysimeter.

A sample collected with a lysimeter may not accurately represent the pore water surrounding the unit due to the sampler location, extraction period, reliance upon suction for sampling, clogging of the porous material, and contamination from the material used to construct the lysimeter. The difficulty of situating a lysimeter in a position where it will intercept water which is characteristic of the bulk soil water is a major problem with lysimetry. Research findings, for example, indicate that more than 90% of the total water flow is channeled through the larger soil pores, while a lysimeter collects water moving primarily in the slower, smaller pore network.[332,333]

The suction and extraction period will determine whether the sample is representative of water moving through the larger or smaller soil pore system. Samples collected in very short periods tend to characterize solutions migrating through the soil macropores held at suctions of 0.1 bars or less.[334] Solutions extracted over longer periods and at higher suctions (≈ 0.1 to 0.45 bar) will characterize the pore water held by the soil near the tension applied by the porous cup which is less susceptible to leaching than the water moving through the macropore system. The investigator must therefore decide which of these two systems is of concern and select the proper suction level and sampling period accordingly.

The use of suction for collecting soil pore water may also contribute to sample bias. Distortion of the normal drainage patterns through accelerated percolation toward the evacuated sampler results in an averaged sample representative of the area surrounding the sampler rather than at that specific depth.[335,336] As extraction continues, flow gradients near the sampler become more pronounced and are measurable within several meters.[337] Flow patterns, especially in the immediate area between the lysimeter and soil, become highly disrupted due partially to the considerable resistance between the porous material of the lysimeter and soil to unsaturated water flow. These flow patterns become less severe with distance from the lysimeter. The resistance of the pore water to suction is also affected in part by the soil structure (e.g., resistance to flow in the micropore system is greater than in the macro system). Because the water chemistry of each system varies to some degree, the source and volume of the pore water entering the lysimeter will influence the chemistry of the composite sample.[338]

Clogging of the porous material occurs in all materials and is a function of unit installation, soil composition, biofilm growth, vacuum application, and soil moisture content. Clogging contributes to additional sorption and screening due to increased exchange and filtering surface areas. Lysimeters with a high free water intake volume (e.g., 530 m ℓ/hr) clog more rapidly than lower intake samplers (265 ,ℓ/hr). Both constant and falling head lysimeters with high intake rates exhibit lower sample concentrations than with low intake rate samplers, presumably due to clogging. While lysimeters with a low intake rate and a falling vacuum appear to be the best combination in terms of reduced clogging, the longer collection period results in increased sample detention time inside the collection vessel. This in turn affects sample representativeness by increasing the possibility of sample redox, pH, and microbiologically induced transformations.

A water sample collected with a lysimeter can also be contaminated by the materials used to construct the unit. Two possible sources are the hydrophilic section and the materials associated with the remainder of the vessel. The degree to which the porous section will contribute potential contaminants or attenuate chemical species to the pore water passing through it is dependent upon the choice of material. The use of solvents for assemblage, rubber stoppers, brass or stainless steel check valves, and inappropriate tubing will present additional opportunities for sample contamination.

a. Ceramic Cup Lysimeters

Ceramic cup lysimeters consist of a plastic tube fitted at one end with a porous cup and plugged by a one or two holed rubber stopper at the opposite end (Figure 1.38).[339-345] The ceramic cup is composed of a porous aluminum oxide with a pore diameter of 2.8 microns. Another design positions the ceramic cup upon the top of a larger diameter PVC chamber.[346] In either configuration, a vacuum is applied through the access tubing; accumulated pore water, which enters the unit via the porous ceramic material, is withdrawn through this tubing to the surface. Subsequent designs for deeper sampling ($\geq$9.1 m) use the two holed stopper to allow separate tubes for chamber evacuation and sample collection through pressurization.

Deep sampling designs are used in cases where suction is inadequate for sample withdrawal. Most of these units incorporate a check valve or separate chamber.[347,348] The check valve arrangement prevents the sample from being forced back through the porous material and into the soil during pressurization, and reduces ceramic fracturing which often occurs under pressure.

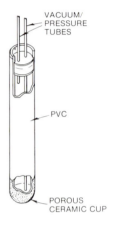

Figure 1.38 Ceramic cup lysimeter with two holed stopper.

Portable or permanent approaches are used for chamber evacuation and sample extraction. For shallow installations, manual extraction and pressurization with a hand pump will suffice or an arrangement with a vacuum chamber similar to Figure 1.39 can be used. A multiple unit approach, in which a mercury pressure control device mounted on the sample flask regulates sample volume, is illustrated in Figure 1.40.[349] This system maintains a constant vacuum within the lysimeter and prevents overfilling of the sample bottle. Other approaches have been designed specifically for volatile organic sampling.[350]

Lysimeters with a ceramic porous cup mounted on a tube require special preparation prior to installation. A recommended procedure includes passing one liter of 8N HCl, followed by rinsing with 15 to 20 liters of distilled water. The unit is considered ready for installation when the specific conductance between cup input and output is less than 2%.[351] Another cleaning procedure involves determining the difference between the dry and wet weight (i.e., in units of pore volume) of the porous material. Approximately 50 to 60 pore volumes of 1N HCl is passed through the material followed by a rinse of 10 pore volumes of deionized water.[352] The unit should soak in distilled water for 15 to 30 minutes prior to installation to ensure saturation of the porous material.

Figure 1.39 Installed lysimeter with sample bottle and vacuum reservoir.[309] (reproduced from EQUIPMENT FOR SOIL RESEARCH, 1981, by permission of the Eijkelkamp, B.V.)

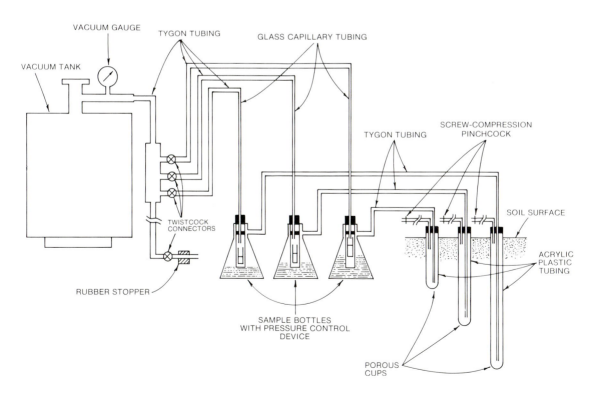

Figure 1.40 Porous cup lysimeters with volume control.[349] (reproduced from SOIL SCIENCE, Volume 124, 1977, page 175, ©1977, Williams and Wilkens Co.)

After excavation of the augered hole, the bottom of the hole is filled with bentonite. A slurry of silica flour or 100 mesh silica sand should be placed above the bentonite. The sampler is forced into the mixture so that the slurry rises several centimeters above the unit. A bentonite seal is placed above the slurry followed by tamped backfill. A third bentonite plug is placed near the soil surface to minimize surface water channeling via the backfilled hole. For multiple installations, an identical procedure is followed for each unit. A locked steel cover casing installed at the surface protects the access tubing.

A number of undesirable characteristics are associated with the use of ceramic as a porous material. The ceramic absorbs or interferes with NO_3-N and P, especially when the sorptive capacity of the soil is less than that of the ceramic.[353,354] Sorption or a combination of factors has been found to reduce levels of Ni, Cu, Pb, Zn, Fe, Mn, and Mg approximately 10% following percolation through ceramic cups.[355] Percolation through the ceramic also reduced concentrations of the chlorinated hydrocarbons ppDDD, ppDDE, and ppDDT by 90%, 70%, and 94% respectively. The porous material can also contribute undesirable substances whose origin is associated with the sintering process. Analysis of extracts from porous ceramic also reveals that Ca, Na, K, Mg, bicarbonate, and silica are leached from the ceramic material.[356]

Researchers have concluded that ceramic does not yield valid water samples for fecal coliform analysis.[357] Other problems include the contribution of contaminants from the solvents used for attaching the porous cup to the PVC chamber, the use of a rubber stopper, and the general fragility of ceramic for field use.

b. Nylon Lysimeters

A nylon mesh has been used as the porous material in a lysimeter. The general configuration, operational characteristics, and installation are identical to the ceramic lysimeter.

A shallow sampling lysimeter with a nylon mesh is illustrated in Figure 1.41. The design of a deep sampling unit is similar. The large pore size of the nylon mesh is suitable only for conditions of high soil moisture and continuous pumping.

c. Fritted Glass Lysimeter

Porous glass has been proposed for the hydrophilic cup of a lysimeter.[359,360] The cup is attached to a PVC tube with a stopper at the opposite end. Installation and operational procedures are identical to the ceramic lysimeter.

d. Teflon® Lysimeters

The development of porous Teflon® tubes and cups provide highly inert material available for the hydrophilic component of a lysimeter. Three sampler designs which utilize porous Teflon® are a shallow sampling, deep sampling, and sleeve lysimeter.

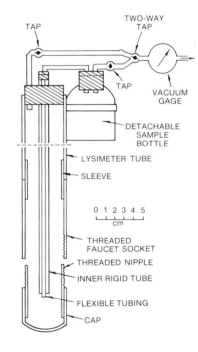

Figure 1.41 Nylon mesh lysimeter.[358] (reproduced from NEW ZEALAND JOURNAL OF SCIENCE, Volume 19, 1976, by permission of the Scientific Information Centre, Wellington, New Zealand)

A shallow sampling Teflon® lysimeter is illustrated in Figure 1.42 and consists of five separate components.[361] These threaded sections are (1) a cap with ferrule connectors for the tubing, (2) blank casing, (3) the porous Teflon® tube, (4) blank casing, and (5) a well point. A threaded eye plug allows the lysimeter to be lowered via a suspension cord rather than the tubing. The modular nature of the design (Figure 1.43) facilitates the interchange of blank casing thereby increasing the volume of sample which can be retained in the vessel.

While most units are constructed with a porous Teflon® section and PVC vessel, the PVC can be replaced with Teflon® when sensitive analysis is required. The use of threaded components also eliminates solvents in the PVC or all Teflon® unit; Teflon® tape is wrapped on the threads prior to joining to ensure that an integral seal is formed between sections.

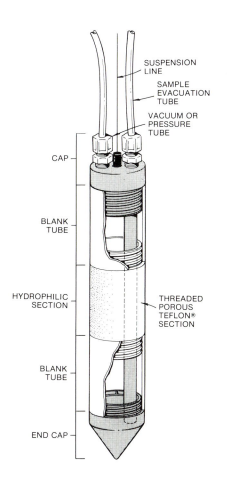

SUSPENSION LINE

SAMPLE EVACUATION TUBE

VACUUM OR PRESSURE TUBE

CAP

BLANK TUBE

HYDROPHILIC SECTION

THREADED POROUS TEFLON® SECTION

BLANK TUBE

END CAP

Figure 1.42 Shallow sampling Teflon® lysimeter. (reproduced from TIMCO GEOTECHNICAL PRODUCTS CATALOGUE, 1982, by permission of Timco Mfg., Inc., Prairie du Sac, WI)

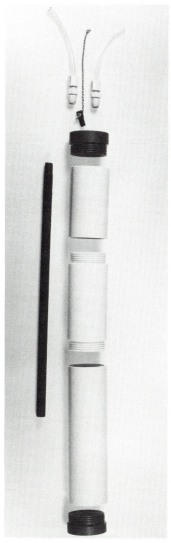

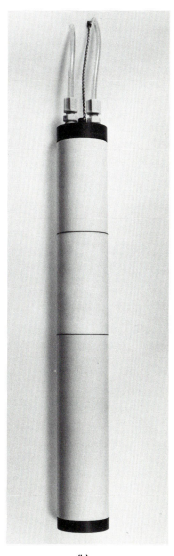

(a)
Disassembled lysimeter

(b)
Assembled lysimeter.

Figure 1.43 Shallow sampling PVC lysimeter with porous Teflon® section. (photographs ©1983, TIMCO MFG., INC., Prairie du Sac, WI)

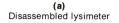

A deep sampling lysimeter (Figure 1.44) allows water entering the vessel to drain to the bottom chamber. The internal check valve isolates this chamber during pressurization. The sample is then brought to the surface. A smaller check valve in the lysimeter cap minimizes sample loss from water film accumulation which can drain back into the lysimeter, thereby biasing the subsequent sample. The sample collecting above the check valve in the cap is withdrawn by pressurizing the vessel several times with a period of 1 to 2 hours between pressurization. For installations greater than 30.5 m, a check valve is frequently placed in the tubing midway between the lysimeter and the surface for the same reason.

A sleeve lysimeter is designed for use in concert with a monitoring well. The sampler surrounds a threaded section of casing which is attached between casing sections. Due to this feature, the unit is functional only for shallow sampling.

The basic components of a sleeve lysimeter and its installation are shown in Figure 1.45. All components are threaded with the same modular features of the deep and shallow sampling Teflon® lysimeters. Decreased chamber volume can therefore be compensated by increasing the length of the lower blank tubing section. An all Teflon® design replaces the PVC portions of the sampler and the internal length of threaded casing when the interference from PVC is a concern.

A difficulty encountered with collecting samples from all lysimeters occurs in cold weather conditions when water film accumulation in the extraction line freezes. Ignoring this problem results in clogging of the extraction line and subsequent unit failure. A bleeder valve (Figure 1.46), developed for Teflon® lysimeters, is installed along the extraction line to eliminate this problem.

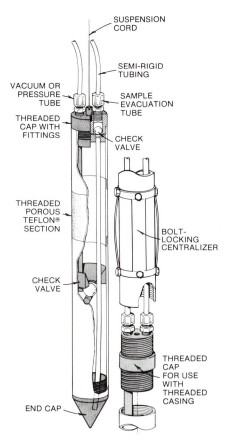

Figure 1.44 Deep sampling Teflon® lysimeter and modified cap. (reproduced from TIMCO GEOTECHNICAL PRODUCTS CATALOGUE, 1982, by permission of Timco Mfg., Inc., Prairie du Sac, WI)

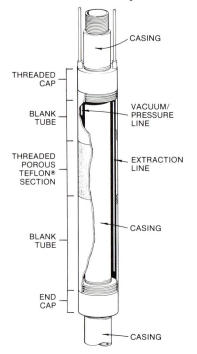

Figure 1.45 Sleeve lysimeter cross section and installation. (courtesy of TIMCO MFG., INC., Prairie du Sac, WI)

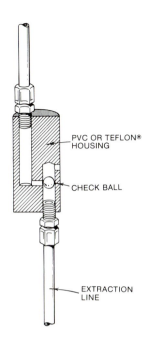

PVC OR TEFLON®
HOUSING

CHECK BALL

EXTRACTION
LINE

Figure 1.46 Cold weather bleeder valve (reproduced from TIMCO GEOTECHNICAL PRODUCTS CATALOGUE, 1982, by permission of Timco Mfg., Inc., Prairie du Sac, WI).

Installation procedures for shallow and deep sampling lysimeters are similar. A hole is excavated at least one meter below where the sampler is to be positioned. A 15 to 20 cm layer of powdered bentonite is placed at the bottom of the hole upon which silica flour slurry is poured. A water to silica flour mixture of 150 mℓ to .45 kg is recommended for the slurry. In order to maximize the effectiveness of this slurry, it is critical to center the lysimeter within the borehole so that the silica flour is uniformly distributed around the unit. A shallow sampler uses the suspension line to center the lysimeter. For a deep sampler, threaded casing with centralizers is threaded onto a specially adapted cap (Figure 1.44b). The casing is left in place for a single sampler or unthreaded and the procedure repeated for multiple installations. The slurry is extended at least 30.5 cm above the lysimeter upon which another layer of powdered bentonite is situated. Another bentonite seal is placed at the surface.

Once the lysimeter is properly installed and backfilled, it is evacuated. A sample, equal to approximately 30% of the volume of water used to mix the silica flour plus the water occupying the pore space of the porous Teflon® from the pre-installation soaking is withdrawn and discarded prior to sample collection. The installation of a sleeve lysimeter (e.g., soaking in distilled water, slurry preparation, bentonite seal placement, etc.) is similar to other Teflon® lysimeters. Because the sleeve lysimeter becomes part of the casing string, care is required to thread the lysimeter onto the casing at a point corresponding to the desired sampling depth when the casing is installed.

A feature inherent in all three Teflon® lysimeter designs is the modularity of the units which allows field assembly of a sampler which is most suited to the soil characteristics in which it is to be installed. Depending upon the soil of interest, the pore size of the hydrophilic section can be selected which best approximates the pore water drainage rate of the soil. Several threaded porous sections in various micron sizes should therefore be available during installation (Figure 1.47).

Figure 1.47 Components and equipment used for the field installation of a shallow sampling Teflon® lysimeter. (photograph ©1983, TIMCO MFG., INC., Prairie du Sac, WI)

51

Teflon® lysimeters do not exhibit many of the water quality problems common with other lysimeters. Leaching experiments with Mg, Mn, Na, Cl, Fe, Pb, and Cd standards show no detectable attenuation of these elements after percolation through the porous Teflon®.[361] Tests with NH_4, PO_4, NO_4, NO_2, and Si standards resulted in a percent recovery within ±4% of the initial concentration.[362] While plugging occurs, it is less than with ceramic presumably due to the more uniform microscoic surface structure of the matrix (Figure 1.48).

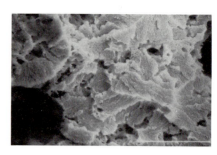

(a) Magnified 700 times.

2. Vacuum Plates and Tubes

Water samples can be collected by evacuation from a porous tube or plate. The principle is similar to lysimetric methods except for the geometry of the porous material.

A vacuum plate consists of an Alundum or ceramic disc attached to an extraction tube. The size of most discs ranges from 4.3 to 25.4 cm in diameter.[363-369] The single tube leading from the disc can be attached to a two way stoppered sample bottle in which a constant vacuum is maintained by a Cartesian monostat and vacuum source at the surface (Figure 1.49). Another arrangement shown in Figure 1.50 uses a 454 ℓ reservoir tank with two control switches for regulating the constant vacuum.

(b) Magnified 5,000 times.

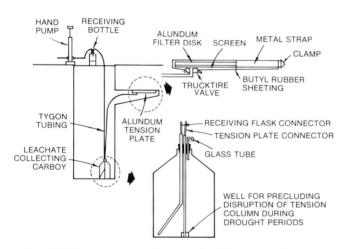

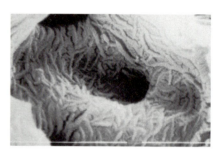

(c) Magnified 15,000 times.

Figure 1.48 Electron scanning photomicrographs of the surface of the porous Teflon®. (photomicrographs ©1983, TIMCO MFG., INC., Prairie du Sac, WI)

Figure 1.49 Alundum vacuum plate installation.[365] (reproduced from SOIL SCIENCE, Volume 85, 1958, page 294, ©1958, Williams and Wilkens Co.)

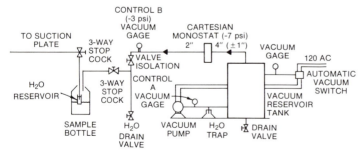

Figure 1.50 Flow diagram for ceramic plate vacuum system.[370]

The typical installation of a vacuum plate requires that a verticle borehole and a second horizontal hole in the side wall is excavated. The second hole should have dimensions similar to that of the plate. Prior to field installation, the plate is cleaned by soaking in a solution of 0.1 N HCl for 24 hours followed by repeated rinses in distilled water. The plates are kept in distilled water until they are inserted into the side wall cavity. Initial and continuous contact between the porous plate and the soil is critical; pneumatic bladders, inner tubes, or similar devices are used to force the plate surface against the ceiling of the side wall excavation.

Ceramic and hollow cellulose fibers, both individual and bundled, have been used for pore water sampling. One design places 12 mm diameter ceramic tubes in a sheet metal trough filled with excavated soil. [371-373] The soil filled trough is placed within a horizontal tunnel and contact between the trough and smooth tunnel ceiling is insured by the inflation of a butyl rubber pillow under the trough. Cellulose acetate fibers used in this manner are thin walled, semipermeable, flexible tubes. An outer diameter of 250 μm is commonly employed.[374] Another type of fiber produced from a noncellulosic polymer has been tested and consists of a 0.1 to 1.5 μm diameter dense inner layer surrounded by a thicker, open celled spongy layer ranging from 50 to 250 μm. The outer structure is more permeable than the inner layer which has a pore size of 0.3 μm.[375] Irrespective of fiber type, all fibers are bundled at one end and sealed. They are then cut to the desired length and the opposite ends attached to a larger PVC or glass tube which acts as a manifold. Suction is applied through tubing from the manifold arrangement. The bundle may be installed directly into the soil, or in a perforated length of PVC tubing which acts as a shield to protect the fibers.

Suction plates collect a large sample volume without extensive disruption of adjacent flow patterns. The effect of extraction time on the soil solution chemistry is also minimal since a large volume of solution is present in the plates before any suction is applied. Ceramic plates exhibit the same potential water quality problems discussed for ceramic cup lysimeters.

Fiber bundles enclosed within a perforated plastic tube can be used to provide an extremely long sampling unit along a horizontal axis. Water quality bias with ceramic, screening of NO_2-N, P, and K by cellulose acetate fibers, and plugging are potential problems.

3. Membrane Filter Samplers

A "Swinnex" type filter holder with a polycarbonate (0.4 to 0.8 μm) or cellulose acetate (0.45 to 5.0 μm) filter and glass fiber prefilter has been proposed for soil pore water sampling (Figure 1.51).[376] The glass fiber prefilter and glass fiber wick covers a membrane filter (Figure 1.51b). Three or four 9 or 11 cm diameter glass fiber sheets are placed at the bottom of a 10 cm hole which are overlaid with two or three smaller glass fiber wick discs. The hole is then backfilled. Polyethylene discs placed at various intervals in the backfill may be used to create layers with decreasing permeability. A sampling tube is connected to the entire unit and is attached to a sample collection bottle at the soil surface (Figure 1.51c). A vacuum of 0.05 to 0.06 bars is recommended.

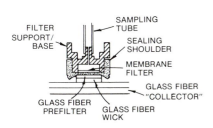

(a) Preparation of "Swinnex" type filter holder for suction sampler.

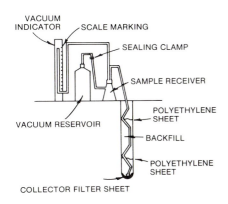

(b) Installation of suction sampler showing glass fiber "wick" and "collector" arrangement.

(c) Installed sampler, sample receiver, and vacuum indicator.

Figure 1.51 "Swinnex" type membrane filter sampler.[376] (reprinted with permission from ENVIRONMENTAL SCIENCE AND TECHNOLOGY, Volume 12, 1978, copyright 1978, American Chemical Society.)

Another approach employs filter paper as a porous media (Figure 1.52). A support and filter paper cemented inside the funnel allows evacuation for sample collection. In this arrangement, a vertical hole and a second horizontal tunnel into the side wall are excavated and the funnel sampling apparatus is placed against the tunnel ceiling.[377] An inflated tire under the funnel insures proper contact between the funnel and soil.

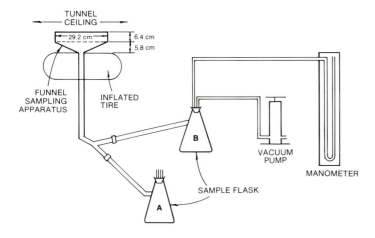

Figure 1.52 Filter membrane funnel sampler.[337] (reproduced from JOURNAL OF ENVIRONMENTAL QUALITY, Volume 8, 1979, page 243, by permission of the American Society of Agronomy)

Membrane filter samplers have many of the limitations inherent with vacuum pressure lysimeters. Another problem unique to membrane filters is the contribution of contaminants from the filter to the filtrate. Depending upon membrane composition and manufacturer, a variety of contaminants may be contributed. Glass fiber filters, for example, contribute nitrogen and carbon particulate matter, as well as sodium, into the filtrate.[378] Thorough membrane rinsing with distilled water prior to installation can minimize these contributions. Biofilm growth on the cellulose acetate membranes can be controlled, in part, with treatments of silver nitrate and sodium chloride.

4. Absorbent Methods

Absorbent methods rely upon the ability of a porous material to absorb soil pore water. A cellulose nylon sponge and ceramic point have been used.

In the former method, a sponge (0.5 x 4.8 x 30 cm) is soaked for 24 hours in a 1 to 5% NaOH solution containing a washing powder. The sponge is then placed within a trough which is positioned against the ceiling of a horizontal tunnel by a series of three lever hinges.[379] When a certain volume of pore water has been absorbed by the sponge, the trough is withdrawn and the sponge is stored in a moisture tight container. Rollers are used to extract the solution from the sponge.

Tapered ceramic rods or points (90 x 12 mm) have been proposed for sampling.[380] The rods are prepared by individual weighing followed by boiling in distilled water, drying for 6 to 7 hours at 105°C, and storage in a disiccator with a saturated NaCl solution in its base. The rods are withdrawn, weighed, and driven into the surface soil. After a certain period of time, the rods are withdrawn and weighed to determine the volume of absorbed water. The points are then leached by boiling them in a known volume of distilled water and the solution analyzed. The original pore water concentration is determined from the ratio of water absorbed by the ceramic to the volume of boiling water.

Absorbent methods are functional primarily at near saturated conditions. The numerous operational difficulties, the potential for sample contamination, and the limited number of parameters which can be tested are major disadvantages with absorbent techniques.

REFERENCES — PART I

1. Chow, V. 1964. Handbook of Applied Hydrology. McGraw-Hill Book Co., N.Y., NY, 1060 p.

2. Childs, E. 1969. An Introduction to the Physical Basis of Soil Water Phenomena. Wiley-Interscience Publ.,N.Y.,NY, 493 p.

3. Hillel, D. 1980. Fundamentals of Soil Physics. Academic Press, N.Y., NY, 413 p.

4. Wilson, L. 1980. Monitoring in the vadose zone: A review of technical elements and methods. U.S. Environmental Protection Agency, EPA-600/7-80, 168 p.

Soil Moisture Potential

5. Hillel, D. 1971. Soil and Water; Physical Principles and Processes. Academic Press, N.Y., NY,288 p.

6. Day, P., G. Bolt, and D. Anderson. 1967. Nature of soil water. R. Hagan, R. Haise, and W. Edminster (Eds.), Irrigation of Agricultural Lands. Agronomy, 11 in series of monographs. Amer. Soc. of Agron., Madison, WI, p.193-208.

7. Topp, G., and W. Zebchuk. 1979. The determination of soil-water desorption curves for soil cores. Can. J. Soil Sci. 59:19-26.

8. Soilmoisture Equipment Corp. 1980. A full range of soilmoisture extractors. Soilmoisture Equipment Corp., Santa Barbara, CA.

Tensiometers

9. Gardner, W., et al. 1922. The capillary potential function and its relation to irrigation practice. Phys. Rev. 20:196.

10. Heck, A. 1934. A soil hygrometer for irrigated cane lands of Hawaii. J. Amer. Soc. Agron. 26:274-278.

11. Rogers, W. 1935. A soil moisture meter depending on the capillary pull of the soil. J. Agric. Sci. 25:326-343.

12. Schofield, R. 1935. The pF of the water in soil. Trans. Third Intl. Cong. Soil. Sci. 2:37-48.

13. Klute, A., and D. Peters. 1962. A recording tensiometer with a short response time. Soil Sci. Soc. Amer. Proc. 26:87-88.

14. Soilmoisture Equipment Corp. 1980. About our tensiometers. Soilmoisture Equipment Corp., Santa Barbara, CA.

15 Timco Mfg., Inc. 1982. Geotechnical Products Catalogue. Timco Mfg., Inc., Prairie du Sac, WI.

16. McKim, H., R. Bert, R. McGaw, R. Atkins, and J. Ingersoll. 1976. Development of a remote-reading tensiometer/transducer system for use in subfreezing temperatures. Proc., Second Conf. on Soil-Water Problems in Cold Regions, Edmonton, Canada, Amer. Geophys. Union, p. 31-45.

17. Wendt, C., O. Wilke, and L. New. 1978. Use of methanol-water solutions for freeze protection of tensiometers. Agron. J. 70:890-891.

18. Dwyer Instruments, Inc. 1981. Dwyer low pressure and control gages. Dwyer Instruments, Inc., Michigan City, IN.

19. Savvides, L., R. Ayers, and M. Ashkar. 1977. A modified mercury tensiometer. Soil Sci. Amer. J. 41:660-661.

20. Perrier, E., and D. Evans. 1961. Soil moisture evaluation by tensiometers. Soil Sci. Soc. Amer. Proc. 25:173-175.

21. Reeve, R. 1965. Hydraulic head. C. Black, D. Evans, J. White, L. Ensimger, and F. Clark (Eds.), Methods of Soil Analyses, Part I. 9 in series monographs. Amer. Soc. of Agron., Madison, WI, p. 180-196.

22. Richards, L. 1949. Methods of measuring soil moisture tension. Soil Sci. 68:95-112.

23. Long, F., and M. Huck. 1980. An automated system for measuring soil water potential gradients in a rhizotron soil profile. Soil Sci. 129:305-310.

24. Long, F. 1982. A new solid-state device for reading tensiometers. Soil Sci. 133:131-132.

25. Leonard, R., and P. Low. 1962. A self-adjusting, null-point tensiometer. Soil Sci. Soc. Amer. Proc. 26:123-125.

26. Watson, K. 1965. Some operating characteristics of a rapid response tensiometer system. Water Resources Res. 1:577-586.

27. Watson, K. 1967. A recording field tensiometer with rapid response characteristics. J. Hydrology 5:33-39.

28. Bianchi, W. 1967. Measuring soil moisture tension changes. Agric. Eng. 43:398-404.

29. Thiel, T., J. Fouss, and A. Leech. 1963. Electrical water pressure transducers for field and laboratory use. Soil Sci. Soc. Amer. Proc. 27:601-602.

30. Thony, J., and G. Vachaud. 1980. Automatic measurement of soil-water pressure using a capacitance manometer. J. Hydrology 46:189-196.

31. Rice, R. 1969. A fast-response, field tensiometer system. Trans. Amer. Soc. Agric. Eng. 12:48-50.

32. Anderson, M., and T. Burt. 1977. Automatic monitoring of soil moisture conditions in a hillslope spur and hollow. J. Hydrology 33:27-36.

33. Burt, T. 1978. An automatic fluid-scanning switch tensiometer system. Brit. Geomorphological Res. Gp. Tech. Bull., p. 1-30.

34. Williams, T. 1978. An automatic scanning and recording tensiometer system. J. Hydrology 39:175-183.

35. Holmes, J., S. Taylor, and S. Richards. 1967. Measurement of soil water. R. Hagan, R. Haise, and W. Edminster (Eds.), Irrigation of Agricultural Lands. Agronomy, 11 in series monographs. Amer. Soc. of Agron., Madison, WI, p. 275 -298.

36. Richards, S. 1965. Soil suction measurements with tensiometers. C. Black, D. Evans, J. White, L. Ensimger, and F. Clark (Eds.), Methods of Soil Analyses, Part I. Agronomy, 9 in series of monographs. Amer. Soc. of Agron., Madison, WI, p. 153-163.

37. Huber, M., and C. Dirksen. 1978. Multiple tensiometer flushing system. Soil Sci. Soc. Amer. J. 42:168-170.

38. Peck, A., and R. Rabbidge. 1966. Direct measurement of moisture potential: A new technique. Proc., UNESCO-Neth. Gov. Symp. Water in the Unsaturated Zone, Wageningen, Netherlands. 1:165-170.

39. Peck, A., and R. Rabbidge. 1966. Soil-water potential: Direct measurement by a new technique. Science 151:1385-1386.

40. Peck, A., and R. Rabbidge. 1969. Design and performance of an osmotic tensiometer for measuring capillary potential. Soil Sci. Soc. Amer. Proc. 33:196-201.

41. Richards, L. 1955. Retention and transmission of water in soil. U.S. Dept. of Agric. Yearbook (1955):144-151.

42. Richards, S., and J. Lamb. 1937. Field measurements of capillary tension. J. Amer. Soc. of Agron. 29:772-780.

43. Schmugge, T., T. Jackson, and H. McKim. 1980. Survey of methods for soil moisture determination. Water Resources Res. 16:961-979.

44. Richards, S., S. Willardson, S. Davis, and J. Spencer. 1973. Tensiometer use in shallow ground water studies. J. Irrigation and Drainage Div., ASCE 99:457-64.

45. Nakano, Y., R. Khalid, and W. Patrick. 1978. Water movement in a land treatment system of wastewater by overland flow. Proc., Intl. Conf. of Wastewater Treatment and Utilization, p. 14/1-14/21.

46. Colbeck, S. 1976. On the use of tensiometers in snow hydrology. J. Glaciology 17:135-140.

47. Wankiewicz, A. 1978. Water pressure in ripe snowpacks. Water Resources Res. 14:593-600.

Thermocouple Psychrometers

48. Barrs, H. 1968. Determination of water deficits in plant tissues. T. Kozlowski (Ed.), Water Deficits and Plant Growth. Academic Press, N.Y., NY, p. 289-300.

49. Spanner, D. 1951. The Peltier effect and its use in the measurement of suction pressure. J. Experimental Botany 2:145-168.

50. Richards, L., and G. Ogata. 1958. Thermocouple for vapor pressure measurement in biological and soil systems at high humidity. Science 128:1089-1090.

51. Richards, L. 1965. Physical condition of water in soil. C. Black, D. Evans, J. White, L. Ensimger, and F. Clark (Eds.), Methods of Soil Analyses, Part I. Agronomy, 9 in series monographs. Amer. Soc. of Agron., Madison, WI, p. 128-152.

52. Peck, A. 1968. Theory of the Spanner psychrometer; 1. The thermocouple. Agric. Meteorol. 5:433-447.

53. Koorevaar, P., and A. Janse. 1972. Some design criteria of thermocouple psychrometers. R. Brown and B. Van Haveren (Eds.), Psychrometry in Water Relations Research. Utah Agric. Experiment Sta., Utah State Univ., Logan, UT, p. 74-83.

54. Dalton, F., and S. Rawlins. 1968. Design criteria for Peltier-effect thermocouple psychrometers. Soil Sci. 105:12-17.

55. Van Haveren, B., and R. Brown. 1972. The properties and behavior of water in the soil-plant-atmosphere continuum. R. Brown and B. Van Haveren (Eds.), Psychrometry in Water Relations Research. Utah Agric. Experiment Sta., Utah State Univ., Logan, UT, p. 1-27.

56. Campbell, G., J. Trull, and W. Gardner. 1968. A welding technique for Peltier thermocouple psychrometers. Soil Sci. Soc. Amer. Proc. 32:887-889.

57. Lopushinsky, W. 1971. An improved welding jig for Peltier thermocouple psychrometers. Soil Sci. Soc. Amer. Proc. 35:149-150.

58. Zanstra, P. 1976. Welding uniform sized thermocouple junctions from thin wires. J. Physics E. Scientific Instruments 9:526-528.

59. Merrill, S., F. Dalton, W. Herkelrath, G. Hoffman, R. Ingvalson, J. Oster, and S. Rawlins. 1968. Details of construction of a multipurpose thermocouple psychrometer. U.S. Salinity Lab., Riverside, CA, ARS-USDA, Res. Rept. 115, p. 9.

60. Brown, R. 1970. Measurement of water potential with thermocouple psychrometers: Construction and applications. U.S. Dept. of Agric. Forest Service, Res. Paper (INT-80), Ogden, UT, p. 27.

61. Lopushinsky, W., and G. Klock. 1971. Construction details of ceramic bulb psychrometers. U.S. Dept. of Agric. Forest Service, Forest Hydrology Lab., Wenatchee, WA, p. 8.

62. Meyn, R., and R. White. 1972. Calibration of thermocouple psychrometers: A suggested procedure for development of a reliable predictive model. R. Brown and B. Van Haveren (Eds.), Psychrometry in Water Relations Research. Utah Agric. Experiment Sta., Utah State Univ., Logan, UT, p. 56-64.

63. Wiebe, H., G. Campbell, W. Gardner, S. Rawlins, J. Cary, and R. Brown. 1971. Measurement of plant and soil water status. Utah State Univ., Logan, UT, Utah Agric. Experiment Sta. Bull. 484, p. 1-71.

64. Merrill, S., and S. Rawlins. 1972. Field measurement of soil water potential with thermocouple psychrometers. Soil Sci. 113: 102-109.

65. Rawlins, S., and F. Dalton. 1967. Psychrometric measurement of soil water potential without precise temperature control. Soil Sci. Soc. Amer. Proc. 31:297-301.

66. Enfield, C., J. Hsieh, and A. Warrick. 1973. Evaluation of water flux above a deep water table using thermocouple psychrometers. Soil Sci. Soc. Amer. Proc. 37:968-970.

67. Brown, R., and R. Johnston. 1976. Extended field use of screen-covered thermocouple psychrometers. Agron. J. 68:995-996.

68. Campbell, E. 1972. Vapor sink and thermal gradient effects on psychrometer calibration. R. Brown and B. Van Haveren (Eds.), Psychrometry in Water Relations Research. Utah Agric. Experiment Sta., Utah State Univ., Logan, UT, p. 94-97.

69. Hoffman, G., J. Oster, and S. Merrill. 1972. Automated measurement of water potential and its components using thermocouple psychrometers. R. Brown and B. Van Haveren (Eds.), Psychrometry in Water Relations Research. Utah Agric. Experiment Sta., Utah State Univ., Logan, UT, p. 123-130.

70. Meeuwig, R. 1972. A low-cost thermocouple psychrometer recording system. R. Brown and B. Van Haveren (Eds.), Psychrometry in Water Relations Research. Utah Agric. Experiment Sta., Utah State Univ., Logan, UT, p. 131-135.

71. Lambert, J., and J. van Schilfgaarde. 1965. A method of determining the water potential of intact plants. Soil Sci. 100:1-9.

72. Hoffman, G., W. Herkelrath, and R. Austin. 1969. Simultaneous cycling of Peltier thermocouple psychrometers for rapid water potential measurements. Agron. J. 61:597-601.

73. Korven, H., and S. Taylor. 1959. The Peltier effect and its use for determining relative activity of soil water. Can. J. of Soil Sci. 39:76-85.

74. Barrs, H., and R. Slayter. 1965. Experience with three vapour methods for measuring water potential in plants. F. Eckardt (Ed.), Proc., Montpellier Symp. on Methodology of Plant Eco-Physiology. UNESCO Arid Zone Research 25:369-384.

75. Hsieh, J., and F. Hungate. 1970. Temperature compensated Peltier psychrometer for measuring plant and soil water potentials. Soil Sci. 110:253-257.

76. Chow, T., and T. deVries. 1973. Dynamic measurement of soil and leaf water potential with a double loop Peltier type thermocouple psychrometer. Soil Sci. Soc. Amer. Proc. 37:181-188.

77. Wiebe, H., R. Brown, and J. Barker. 1977. Temperature gradient effects on in situ hygrometer measurements of water potential. Agron. J. 69:933-939.

78. Ingvalson, R., J. Oster, S. Rawlins, and G. Hoffman. 1970. Measurement of water potential and osmotic potential in soil with a combined thermocouple psychrometer and salinity sensor. Soil Sci. Soc. Amer. Proc. 34:570-574.

79. Monteith, J., and P. Owen. 1958. A thermocouple method for measuring relative humidity in the range 95-100%. J. Scientific Instruments 35:433-446.

80. Box, J. 1965. Design and calibration of a thermocouple psychrometer which uses the Peltier effect. A. Weiler (Ed.), Humidity and Moisture. Reinhold Publishing Corp., N.Y., NY, I:110-121.

81. Rawlins, S. 1966. Theory for thermocouple psychrometers used to measure water potential in soil and plant samples. Agric. Meteorol. 3:293-310.

82. Lang, A., and E. Trickett. 1965. Automatic scanning of Spanner and droplet psychrometers having outputs up to $30\,\mu V$. J. Scientific Instruments 42:777-782.

83. Lang, A. 1967. Osmotic coefficients and water potentials of sodium chloride solutions from 0 to 40°C. Austral. J. Chem. 20:2017-2023.

84. Campbell, G., and W. Gardner. 1971. Psychrometric measurement of soil water potential: Temperature and bulk density effects. Soil Sci. Soc. Amer. Proc. 35:8-12.

85. Zollinger, W., G. Campbell, and S. Taylor. 1966. A comparison of water-potential measurements made using two types of thermocouple psychrometer. Soil Sci. 102:231-239.

86. Brown, R. and J. Collins. 1980. A screen-caged thermocouple psychrometer and calibration chamber for measurements of plant and soil water potential. Agron. J. 72:851-853.

87. Moore, R., and M. Caldwell. 1972. The field use of thermocouple psychrometers in desert soils. R. Brown and B. Van Haveren (Eds.), Psychrometry in Water Relations Research. Utah Agric. Experiment Sta., Utah State Univ., Logan, UT, p. 165-169.

88. Hsieh, J., C. Enfield, and F. Hungate. 1972. Application of temperature-compensated psychrometers to the measurement of water potential gradients. R. Brown and B. Van Haveren (Eds.), Psychrometry in Water Relations Research. Utah Agric. Experiment Sta., Utah State Univ., Logan, UT, p. 154-158.

89. Van Haveren, B. 1972. Measurements of relative vapor pressure in snow with thermocouple psychrometers. R. Brown and B. Van Haveren (Eds.), Psychrometry in Water Relations Research. Utah Agric. Experiment Sta., Utah State Univ., Logan, UT, p. 178-185.

90. Campbell, G. 1979. Improved thermocouple psychrometers for measurement of soil water potential in a temperature gradient. J. Physics E. Scientific Instruments 12:739-743.

91. Daniel, D., J. Hamilton, and R. Olson. 1981. Suitability of thermocouple psychrometers for studying moisture movement in unsaturated soils. T. Zimmie and C. Riggs (Eds.), Permeability and Groundwater Contaminant Transport. Amer. Soc. for Testing and Materials, Philadelphia, PA, ASTM STP 746, p. 84-100.

Laboratory Methods

92. Gardner, W. 1965. Water content. C. Black, D. Evans, J. White, L. Ensimger, and F. Clark (Eds.), Methods of Soil Analyses, Part I. Agronomy, 9 in series monographs. Amer. Soc. of Agron., Madison, WI, p. 82-125.

93. Reynolds, S. 1970. The gravimetric method of soil moisture determiniation, Part I: A study of equipment, and methodological problems. J. Hydrology 11:258-273.

94. Reynolds, S. 1970. The gravimetric method of soil moisture determination, Part III: An examination of factors influencing soil moisture variability. J. Hydrology 11:288-300.

95. Holtan, H., N. Minshall, and L. Harrold. 1962. Field manual for research agricultural hydrology. U.S. Agric. Res. Service, Soil and Water Conservation Res. Div. Agric. Handbook 224, 215 p.

96. Geary, P. 1956. Determination of moisture in solids. Brit. Sci. Instrument Res. Assn., Res. Rept. M.24:52.

97. Thijssen, H., C. de Witt, E. vanVollenhoven, H. Timmers, and L. Admiraal. 1954. New instruments for agricultural research. Neth. J. Agric. Res. 2:209-214.

98. Bouyoucos, G. 1928. Determining soil moisture rapidly and accurately by methyl alcohol. J. Amer. Soc. of Agron. 20:82-83.

99. Bouyoucos, G. 1937. Evaporating the water with burning alcohol as a rapid means of determining moisture content of soils. Soil Sci., 44:377-383.

100. Blystone, J., A. Pelzner, and G. Steffens. 1961. Moisture content determination by the calcium carbide gas pressure method. Public Roads 31:177-181.

101. Ballard, L. 1973. Instrumentation for measurement of moisture: Literature review and recommended research. National Cooperative Highway Res. Program Rept. 138, 60 p.

102. Fraade, D. 1963. Measuring moisture in solids. Instruments and Control Systems 36:99-101.

103. Wilde, S., and D. Spyridakis. 1962. Determination of soil moisture by the immersion method. Soil Sci. 94:132-133.

104. Garton, J., and F. Crow. 1954. Rapid methods of determining soil moisture. Agric. Eng. 35:486-487, 491.

105. Hancock, C., and R. Burdick. 1957. Rapid determination of water in wet soils. Soil Sci. 83:197-205.

106. Bottcher, C. 1952. Theory of Electric Polarisation. Elsevier Publishing Co., Amsterdam, Netherlands, 492 p.

Soil Moisture Blocks

107. Haise, H., and O. Kelley. 1946. Relation of moisture tension to heat transfer and electrical resistance in plaster of paris blocks. Soil Sci. 61:411-422.

108. Bouyoucos, G., and A. Mick. 1948. A fabric absorption unit for continuous measurement of soil moisture in the field. Soil Sci. 66:217-232.

109. Bouyoucos, G. 1952. Methods for measuring the moisture content of soils under field conditions. Frost Action in Soils, A Symp. Highway Res. Board, Special Rept. No. 2. National Res. Council Publ. 213, p. 64-74.

110. Bouyoucos, G. 1954. New type electrode for plaster of paris moisture blocks. Soil Sci. 78:339-342.

111. Soilmoisture Equipment Corp. 1974. Model no. 5200 soilmoisture blocks—Bull. A14. Soilmoisture Equipment Corp., Santa Barbara, CA.

112. Slater, C. 1942. A modified resistance block for soil moisture measurements. J. Amer. Soc. Agron. 34:284-285.

113. Pereira, H. 1951. A cylindrical gypsum block for moisture studies in deep soils. J. Soil Sci. 2:212-223.

114. Croney, D., J. Coleman, and E. Currier. 1951. The electrical resistance method of measuring soil moisture. Brit. J. Appl. Phys. 2:85-91.

115. Anderson, A., and N. Edlefsen. 1942. Laboratory study of the response of 2- and 4-electrode plaster of paris blocks as soil moisture indicators. Soil Sci. 53:413-428.

116. Bouyoucos, G., and A. Mick. 1940. An electrical resistance method for the continuous measurement of soil moisture under field conditions. Mich. Agric. Experiment Sta. Tech. Bull. 172:3-38.

117. Michelson, L., and W. Lord. 1962. The use and construction of concentric gypsum soil moisture sensing units and a rapid method of determining mean resistance and moisture values. Amer. Soc. Hort. Sci. 81:565-567.

118. Kemper, W., and M. Amemiya. 1958. Utilization of air permeability of porous ceramics as a measure of hydraulic stress in soils. Soil Sci. 85:117-124.

119. Becker, J., G. Green, and G. Pearson. 1946. Properties and use of thermistors-thermally sensitive resistors. Elec. Eng. Trans. 65:711-725.

120. Colman, E., and T. Hendrix. 1949. Fiberglass electrical soil-moisture instrument. Soil Sci. 67:425-438.

121. Bouyoucos, G. 1949. Nylon electrical resistance unit for continuous measurement of soil moisture in the field. Soil Sci. 67:319-330.

122. Hancox, N., and J. Walker. 1966. The influence of liquid resistivity changes on plaster of paris resistance and capacitance moisture gauges. Brit. J. Appl. Phys. 17:827-833.

123. Bourget, S., D. Elrick, and C. Tanner. 1958. Electrical resistance units for moisture measurements: Their moisture hysteresis, uniformity, and sensitivity. Soil Sci. 86:298-304.

124. Phene, C., G. Hoffman, and S. Rawlins. 1971. Measuring soil matric potential in situ by sensing heat dissipation within a porous body: I. Theory and sensor construction. Soil Sci. Soc. Amer. Proc. 35:27-33.

125. Taylor, S. 1955. Field determinations of soil moisture. Agric. Eng. 36:654-659.

126. Postlethwaite, J., and E. Trickett. 1956. The measurement of soil moisture. J. Agric. Eng. Res. 1:89-95.

127. El-Samie, A., and A. Marsh. 1955. A tube containing gypsum blocks for following moisture changes in undisturbed soil. Soil Sci. Soc. Amer. Proc. 19:404-406.

128. Perrier, E., and A. Marsh. 1958. Performance characteristics of various electrical resistance units and gypsum materials. Soil Sci. 86:140-147.

129. Salaruddin, M., and B. Khasbardar. 1967. An instrument for soil moisture determination. Indian J. Tech. 5:296-299.

130. Schlub, R., and J. Maine. 1979. Portable recorder for the continuous monitoring of soil moisture resistance blocks. J. Agric. Eng. Res. 24:319-323.

131. Williams, T. 1980. An automatic electrical resistance soil-moisture measuring system. J. Hydrology 46:385-390.

132. Aitchison, G., and P. Butler. 1951. Gypsum block moisture meters as instruments for the measurement of tension in water. Austral. J. Appl. Sci. 2:257-266.

133. Shaw, B., and L. Baver. 1939. An electrothermal method for following moisture changes of the soil in situ. Soil Sci. Soc. Amer. Proc. 4:78-83.

134. Kelley, O. 1944. A rapid method of calibrating various instruments for measuring soil moisture in situ. Soil Sci. 58:433-440.

135. Richards, L., and L. Weaver. 1943. The sorption-block soil moisture meter and hysteresis effects related to its operation. J. Amer. Soc. Agron. 35:1002-1011.

Four Electrode Method

136. Edlefsen, N., and A. Anderson. 1941. The four-electrode resistance method for measuring soil moisture content under field conditions. Soil Sci. 51:367-376.

137. Vasa, J. 1969. Some methods for the determination of soil moisture and balance measuring. Proc., UNESCO-Neth. Gov. Symp. Water in the Unsaturated Zone. Wageningen, Netherlands, 1:119-124.

138. Kirkham, D., and G. Taylor. 1949. Some tests on a four-electrode probe for soil moisture measurement. Soil Sci. Soc. Amer. Proc. 14:42-46.

139. Bally, R. 1969. Determination of the coefficients of water migration through soils. Proc., UNESCO-Neth. Gov. Symp. Water in the Unsaturated Zone. Wageningen, Netherlands, 1:245-256.

140. Bunnenberg, C., and W. Kuhn. 1980. An electrical conductance method for determining condensation and evaporation processes in arid soils with high spatial resolution. Soil Sci. 129:58-66.

Capacitance Sensors

141. Selig, E., D. Wobschall, S. Mansukhani, and A. Motiwala. 1975. Capacitance sensor for soil moisture measurement. Frost, Moisture and Erosion. Transportation Res. Bd., Transportation Res. Record 532, p. 64-75.

142. Birchak, J., C. Gardner, and H. Hipp. 1974. High dielectric constant microwave probes for sensing soil moisture. Proc., IEEE 62:93-98.

143. Thomas, A. 1966. In situ measurement of moisture in soil and similar substances by "fringe" capacitance. J. Scientific Instruments 43:21-27.

144. Wobschall, D. 1978. A frequency shift dielectric soil moisture sensor. IEEE Trans. Geoscience Electronics GE-16:112-118.

145. McKim, H., R. Layman, J. Walsh, and T. Pangburn. 1979. Comparison of radio frequency, tensiometer and gravimetric soil moisture techniques. Plasma Phys. Lab. Rept., Dartmouth College, Hanover, NH, p. 129-135.

146. Walsh, J., D. McQueeney, R. Layman, and H. McKim. 1979, Development of a simplified method for field monitoring of soil moisture. Proc., 2nd Colloquium on Planetary Water and Polar Processes, Hanover, NH, p. 40-44.

147. Matthews, J. 1963. The design of an electrical capacitance-type moisture meter for agricultural use. J. Agric. Eng. Res. 8:17-30.

148. Matzkanin, G., E. Selig, and D. Wobschall. 1979. Instrumentation for moisture measurement-bases, subgrades, and earth materials (sensor evaluation). Transportation Res. Board, NCHRP 21-2(3). p. 41.

149. Kuráž, V. 1981. Testing of a field dielectric soil moisture meter. ASTM Geotechnical Testing J. 4:111-116.

150. Kuráž, V., M. Kutilek, and I. Kaspar. 1970. Resonance-capacitance soil moisture meter. Soil Sci. 110:278-279.

151. Kuráž, V., and J. Matousek. 1977. A new dielectric soil moisture meter for field measurement of soil moisture. Intl. Commission on Irrigation and Drainage Bull. 26:76-79.

152. Silva, L., F. Schultz, and J. Zalusky. 1974. Electrical methods of determining soil moisture content. Lab. for Applications of Remote Sensing, Purdue Univ., Lafayette, IN, Information Note 112174, p. 165.

153. Cihlar, J., and F. Ulaby. 1974. Dielectric properties of soils as a function of moisture content. Univ. of Kansas Center for Research, Inc.,Lawrence, KS, RSL Tech. Rept. No. 177-47, p. 61.

154. Mack, A., and E. Brach. 1966. Soil moisture measurement with ultrasonic energy. Soil Sci. Soc. Amer. Proc. 30:544-548.

155. Hoekstra, P., and A. Delaney. 1974. Dielectric properties of soils at UHF and microwave frequencies. J. Geophysical Res. 79:1699-1708.

156. Wang, J., and T. Schmugge. 1978. An empirical model for the complex dielectric permittivity of soils as a function of water content. NASA Tech. Memo. 79659, p. 33.

Electrothermal Methods

157. Aldous, W., and W. Lawton. 1952. The measurement of soil moisture and temperature by heat diffusion type moisture cell. Frost Action in Soils, A Symp. Highway Res. Board, Special Rept. No. 2. National Res. Council Publ. 213, p. 74-95.

158. Phene, C., S. Rawlins, and G. Hoffman. 1971. Measuring soil matric potential in situ by sensing heat dissipation within a porous body: II. Experimental results. Soil Sci. Soc. Amer. Proc. 35:225-229.

159. Phene, C., G. Hoffman, and R. Austin. 1973. Controlling automated irrigation with soil matric potential sensor. Trans. ASAE 16:773-776.

160. Bloodworth, M., and J. Page. 1957. Use of thermistors for the measurement of soil moisture and temperature. Soil Sci. Soc. Amer. Proc. 21:11-15.

161. Philip, J. 1961. The theory of heat flux meters. J. Geophys. Res. 66:571-579.

162. Bloomer, J., and J. Ward. 1979. A semi-automatic field apparatus for the measurement of thermal conductivities of sedimentary rocks. J. Physics E. Scientific Instruments 12:1033-1035.

163. Shaw, B., and L. Baver. 1939. An electrothermal method for following moisture changes of the soil in situ. Soil Sci. Soc. Amer. Proc. 4:78-83.

164. DeVries, D. 1952. A nonstationary method for determining thermal conductivity of soil in situ. Soil Sci. 73:83-89.

165. Fuchs, M., and C. Tanner. 1968. Calibration and field test of soil heat flux plates. Soil Sci. Soc. Amer. Proc. 32:326-328.

166. Fuchs, M., and A. Hadas. 1973. Analysis of the performance of an improved soil heat flux transducer. Soil Sci. Soc. Amer. Proc. 37:173-175.

167. Shaw, B., and L. Baver. 1939. Heat conductivity as an index of soil moisture. J. Amer. Soc. Agron. 31:866-891.

168. DeVries, D., and A. Peck. 1958. On the cylindrical probe method of measuring thermal conductivity with special reference to soils: I. Extension of theory and discussion of probe characteristics. Austral. J. Phys. 11:225-271.

169. Momin, A. 1945. A new simple method of estimating the moisture content of soil in situ. Indian J. of Agric. Sci. 17:81-85.

170. Kubo, J. 1953. A new method for the soil moisture measurement. J. Agric. Meteorol. of Tokyo 8:108-110.

171. Hooper, F. 1952. The thermal conductivity probe. Frost Action in Soils, A Symp. Highway Res. Board, Special Rept. No. 2. National Res. Council Publ. 213, p. 57-59.

172. Wechsler, A., P. Glaser, and R. McConnell. 1965. Methods of laboratory and field measurements of thermal conductivity of soils. Cold Regions Res. and Eng. Lab., Hanover, NH, Special Rept. 82, p. 31.

173. Fritton, D., W. Busscher, and J. Alpert. 1974. An inexpensive but durable thermal conductivity probe for field use. Soil Sci. Soc. Amer. Proc. 38:854-855.

174. Beck, A., F. Anglin, and J. Sass. 1971. Analysis of heat flow data — in situ thermal conductivity measurements. Can. J. Earth Sci. 8:1-19.

175. Slusarchuk, W., and P. Fougler. 1973. Development and calibration of a thermal conductivity probe apparatus for use in the field and laboratory. Natural Resource Council of Canada, Div. of Building Res., TP388, NRCC13267, p. 18.

176. Sophocleous, M. 1978. Analysis of heat and water transport in unsaturated-saturated porous media. Ph.D. Thesis. Univ. of Alberta, Edmonton, Alberta, Canada, p. 271.

177. Sophocleous, M. 1979. A thermal conductivity probe designed for easy installation and recovery from shallow depths. Soil Sci. Soc. Amer. J. 43:1056-1058.

178. Hooper, F., and F. Leeper. 1950. Transit heat flow apparatus for the determination of thermal conductivities. Trans. Amer. Soc. of Heating and Ventilation Eng. 56:309-329.

179. DeVries, D. 1953. Some results of field determinations of the moisture content of soil from thermal conductivity measurements. Neth. J. Agric. Sci. 1:115-121.

180. Van Duin, R., and D. DeVries. 1954. A recording apparatus for measuring thermal conductivity, and some results obtained with it in soil. Neth. J. Agric. Sci. 2:166-175.

181. Blackwell, J. 1954. A transient-flow method for determination of thermal constants of insulating materials in bulk: Part I. Theory. J. Appl. Phys. 25:137-144.

182. Blackwell, J. 1956. The axial flow error in the thermal conductivity probe. Can. J. Phys. 34:412-417.

183. Mann, G., and F. Forsyth. 1956. Measurement of the conductivity of samples of thermal insulating material and of insulation in situ by the water probe method. Modern Refrigerator Air Control 59:188-191.

184. Jaeger, J. 1958. The measurement of thermal conductivity and diffusivity with cylindrical probes. Trans. Amer. Geophys. Union 39:708-710.

185. DeVries, D., and A. Peck. 1958. On the cylindrical probe method of measuring thermal conductivity with special reference to soils: II. Analysis of moisture effects. Austral. J. Phys. 11:409-423.

186. Carslaw, H., and J. Jaeger. 1960. Conduction of Heat in Solids. Oxford Univ. Press, London, England, 350 p.

187. Jackson, R., and S. Taylor. 1965. Heat transfer. C. Black, D. Evans, J. White, L. Ensimger,and F. Clark (Eds.), Methods of Soil Analyses, Part I. Agronomy, 9 in series monographs. Amer. Soc. of Agron., Madison, WI, p. 349-360.

188. Overgaard, M. 1970. The calibration factor of heat flux meters in relation to the thermal conductivity of the surrounding medium. Agric. Meteorol. 7:401-410.

189. Cummings, R., and R. Chandler. 1940. A field comparison of the electrothermal and gypsum block electrical resistance methods with the tensiometer method for estimating soil moisture in situ. Soil Sci. Soc. Amer. Proc. 5:80-85.

Neutron Thermalization

190. Stone, J., R. Shaw, and D. Kirkham. 1960. Statistical parameters and reproducibility of the neutron method for measuring soil moisture. Soil Sci. Soc. Amer. Proc. 24:435-438.

191. van Bavel, C. 1958. Measurement of soil moisture content by the neutron method. U.S. Dept. of Agric., Agric. Res. Service ARS41-24:1-29.

192. van Bavel, C., and G. Stirk. 1967. Soil water measurement with an Am[241]-Be neutron source and an application to evaporimetry. J. Hydrology 5:40-60.

193. McHenry, J. 1963. Theory and application of neutron scattering in the measurement of soil moisture. Soil Sci. 95:294-307.

194. Bell, J., and J. McCullock. 1966. Soil moisture estimation by the neutron scattering method in Britain. J. Hydrology 4:254-263.

195. Long, I., and B. French. 1967. Measurement of soil moisture in the field by neutron moderation. J. Soil Sci. 18:149-166.

196. Bell, J. 1969. A new design principle for neutron soil moisture gages: The "Wallingford" neutron probe. Soil Sci. 108:160-164.

197. Parkes, M., and N. Siam. 1979. Error associated with measurement of soil moisture change by neutron probe. J. Agric. Eng. Res. 24:87-93.

198. van Bavel, C. 1962. Accuracy and source strength in soil moisture neutron probes. Soil Sci. Soc. Amer. Proc. 26:405.

199. Lawless, G., N. MacGillivray, and P. Nixon. 1963. Soil moisture interface effects upon readings of neutron moisture probes. Soil Sci. Soc. Amer. Proc. 27:502-507.

200. Pierpoint, G. 1966. Measuring surface soil moisture with the neutron depth probe and surface shield. Soil Sci. 101:189-192.

201. Black, J., and P. Mitchell. 1968. Near surface soil moisture measurement with a neutron probe. J. Austral. Inst. of Agric. Sci. 34:181-182.

202. Belcher, D. 1952. The measurement of soil moisture and density by neutron and gamma-ray scattering. Frost Action in Soils, A Symp. Highway Res. Board, Special Rept. No. 2. National Res. Council Publ. 213, p. 98-110.

203. Stone, J., D. Kirkham, and A. Read. 1955. Soil moisture determination by a portable neutron scattering moisture meter. Soil Sci. Soc. Amer. Proc. 19:419-423.

204. van Bavel, C., and N. Underwood. 1956. Neutron and gamma radiation as applied to measuring physical properties of soil in its natural state. Trans. Intl. Cong. of Soil Sci., 6th Cong., Paris, France B:355-360.

205. Holmes, J., and K. Turner. 1958. The measurement of water content of soils by neutron scattering: A portable apparatus for field use. J. Agric. Eng. Res. 3:199-204.

206. van Bavel, C., D. Nielsen, and J. Davidson. 1961. Calibration and characteristics of two neutron moisture probes. Soil Sci. Soc. Amer. Proc. 25: 329-334.

207. DeVries, J., and K. King. 1961. Note on the volume of influence of a neutron surface moisture probe. Can. J. Soil Sci. 41:253-257.

208. Belcher, D., T. Cuykendall, and H. Sack. 1952. Nuclear methods for measuring soil density and moisture in thin soil layers. Civil Aeronautics Admin., Tech. Devel. Rept. No. 161.

209. Phillips, R., C. Jensen, and D. Kirkham. 1960. Use of radiation equipment for plow-layer density and moisture. Soil Sci. 89:2-7.

210. van Bavel, C. 1961. Neutron measurement of surface soil moisture. J. Geophys. Res. 66:1493-1498.

211. Cope, F., and E. Trickett. 1965. Measuring soil moisture. Soil and Fertilizers 28:201-208.

212. Visvalingam, M., and J. Tandy. 1972. The neutron method for measuring soil moisture content—A review. J. Soil Sci. 23:499-511.

213. Glenn, D., R. Henderson, and F. Bolton. 1980. A retractable, neutron-probe access tube. Agron. J. 72:1067-1068.

214. Hanks, R., and S. Bowers. 1960. Neutron meter access tube influences soil temperature. Soil Sci. Soc. Amer. Proc. 24:62-63.

215. Stolzy, L., and G. Cahoon. 1957. A field-calibrated portable neutron rate meter for measuring soil moisture in citrus orchards. Soil Sci. Soc. Amer. Proc. 21:571-575.

216. Weinberg, A., and E. Wigner. 1958. The Physical Theory of Neutron Chain Reactors. Univ. of Chicago Press, Chicago, IL, 800 p.

217. Shirazi, G., and M. Isobe. 1976. Calibration of neutron probe in some selected Hawaiian soils. Soil Sci. 122:165-170.

218. Nakayama, F., and R. Reginato. 1982. Simplifying neutron moisture meter calibration. Soil Sci. 133:48-52.

219. Rawls, W., and L. Asmussen. 1973. Neutron probe field calibration for soils in the Georgia coastal plain. Soil Sci. 116:262-265.

220. Bell, J., and C. Eeles. 1967. Neutron random counting error in terms of soil moisture for nonlinear calibration curves. Soil Sci. 103:1-3.

221. Ursic, S. 1967. Improved standards for neutron soil water meters. Soil Sci. 104:323-325.

222. Stirk, G. 1972. Moisture measurement in swelling clay soils. Symp. on Physical Aspects of Swelling Clay Soils, Univ. of New England, p. 53-61.

223. Greacen, E., and G. Schrale. 1976. The effect of bulk density on neutron meter calibration. Austral. J. Soil Res. 14:159-169.

224. Greacen, E., and C. Hignett. 1979. Sources of bias in the field calibration of a neutron meter. Austral. J. Soil Res. 17:405-415.

225. Johnson, A. 1962. Methods of measuring soil moisture in the field. U.S. Geological Survey, Water-Supply Paper 1619-U, p. 25.

226. Gardner, W., and D. Kirkham. 1952. Determination of soil moisture by neutron scattering. Soil Sci. 73:391-401.

227. Brakensiek, D., H. Osborn, and W. Rawls. 1979. Field manual for research in agricultural hydrology. Sciences and Education Admin., U.S. Dept. of Agric., Agric. Handbook 224: 505-547.

228. Holmes, J., and A. Jenkinson. 1959. Techniques for using the neutron moisture meter. J. Agric. Eng. Res. 4:100-109.

Gamma Ray Attenuation

229. Ryhiner, A., and J. Pankow. 1969. Soil moisture measurement by the gamma transmission method. J. Hydrology 9:194-205.

230. Gardner, W., G. Campbell, and C. Calissendorff. 1969. Water content and soil bulk density measured concurrently using two gamma photon energies. Washington State Univ., Pullman, WA, U.S. A.E.C. Rept. RLO-1543-6, p. 42.

231. Gardner, W., G. Campbell, and C. Calissendorff. 1972. Systematic and random errors in dual gamma energy soil bulk density and water content measurements. Soil Sci. Amer. Proc. 36:393-398.

232. Reginato, R., and K. Stout. 1970. Temperature stabilization of gamma ray transmission equipment. Soil Sci. Amer. Proc. 34:152-153.

233. Bouwer, H., and R. Jackson. 1974. Determining soil properties. J. van Schilfgaarde, R. Dinauer, and M. Davis (Eds.), Drainage for Agriculture. Agronomy, 17 in series monographs. Amer. Soc. of Agron., Madison, WI, p. 611-675.

234. Soane, B., and J. Henshall. 1979. Spatial resolution and calibration characteristics of two narrow probe gamma-ray transmission systems for the measurement of soil bulk density in situ. J. Soil Sci. 30:517-528.

235. Corey, J., S. Peterson, and M. Wakat. 1971. Measurement of attenuation of ^{137}Cs and ^{241}Am gamma rays for soil density and water content determinations. Soil Sci. Amer. Proc. 35: 215-219.

236. Keys, W., and L. MacCary. 1971. Application of borehole geophysics to water resources investigations. Techniques of Water-Resources Investigations of the United States Geological Survey: Application of Borehole Geophysics to Water-Resources Investigations, Book 2, Chapter E1, p. 70-74.

237. Couchat, P., P. Moutonnet, and M. Puard. 1979. The application of the gamma neutron method for transport studies in field soils. Water Resources Res. 15:1583-1588.

238. Belcher, D., T. Cuykendall, and H. Sach. 1950. The measurement of soil moisture and density by neutron and gamma-ray scattering. Civil Aeronautics Admin., Tech. Devel. Rept. No. 127, p. 1-20.

239. Soane, B. 1967. Dual energy gamma-ray transmission for coincident measurement of water content and dry bulk density of soil. Nature 214: 1273-1274.

240. Soane, B. 1967. Double energy gamma ray transmission for moisture and density measurement in soil tillage studies. Intl. Soil Water Symp., Prague, Czechoslavakia, p. 197.

241. Nofziger, D., and D. Swartzendruber. 1974. Material content of binary physical mixtures as measured with a dual-energy beam of γ-rays. J. Appl. Phys. 45:5443-5449.

242. Nofziger, D. 1978. Errors in gamma-ray measurements of water content and bulk density in nonuniform soils. Soil Sci. Soc. Amer. J. 42: 845-850.

243. Goit, J., P. Groenevelt, B. Kay, and J. Loch. 1978. The applicability of dual gamma scanning to freezing soils and the problem of stratification. Soil Sci. Soc. Amer. J. 42:858-863.

244. Wood, B., and N. Collis-George. 1980. Moisture content and bulk density measurements using dual-energy beam of gamma radiation. Soil Sci. Soc. Amer. J. 44:662-663.

245. Vomocil, J. 1954. In situ measurement of soil bulk density. Agric. Eng. 35:651-654.

246. van Bavel, C., N. Underwood, and S. Ragar. 1957. Transmission of gamma radiation by soils and soil densitometry. Soil Sci. Soc. Amer. Proc. 21: 588-591.

247. Ferguson, J., and W. Gardner. 1962. Water content measurement in soil columns by gamma ray absorbtion. Soil Sci. Soc. Amer. Proc. 26:11-18.

248. Gurr, C. 1962. Use of gamma rays in measuring water content and permeability in unsaturated columns of soil. Soil Sci. 94:224-229.

249. Gardner, W. 1965. Water content. C. Black, D. Evans, J. White, L. Ensimger, and F. Clark (Eds.), Methods of Soil Analyses, Part I. Agronomy, 9 in series monographs. Amer. Soc. of Agron., Madison, WI, p. 82-125.

250. Rockwell, T. (Ed.). 1956. Reactor Shielding Design Manual. McGraw-Hill Book Co., N.Y., NY, 481 p.

251. Kozachyn, J., and J. McHenry. 1964. A method for installation of access tubes and the development of field equipment for measuring soil moisture by neutron scatter. Agron. J. 56:443-444.

252. Myhre, D., J. Sanford, and W. Jones. 1969. Apparatus and technique for installing access tubes in soil profiles to measure soil water. Soil Sci. 108:296-299.

253. Reginato, R., and R. Jackson. 1971. Field measurement of soil water content by gamma-ray transmission compensated for temperature fluctuations. Soil Sci. Soc. Amer. Proc. 35:529-533.

Nuclear Magnetic Resonance

254. Andreyev, S., and B. Martens. 1960. Soil moisture determination by the method of nuclear magnetic resonance. Soviet Soil Sci. (Pochvovedenie) 10:1129-1132.

255. Graham, J., G. Walter, and G. West. 1964. Nuclear magnetic resonance study of interlayer water in hydrated layer silicates. J. Chem. Phys. 40:540-550.

256. Wu, T. 1964. A nuclear magnetic resonance study of water in clay. J. Geophys. Res. 69:1083-1091.

257. Woessner, D., and B. Snowden. 1969. A study of the orientation of adsorbed water molecules on montmorillonite clays by the pulsed NMR. J. Colloid. Interface Sci. 30:54-68.

258. Prebble, R., and J. Currie. 1970. Soil water measurement by a low-resolution nuclear magnetic resonance technique. J. Soil Sci. 21:273-288.

259. Pearson, R., and L. Derbyshire. 1973. NMR studies of water adsorbed on a number of silica surfaces. J. Colloid. Interface Sci. 46:232-248.

260. Tice, A., C. Burrous, and D. Anderson. 1978. Phase composition measurement on soils at very high water contents by the pulsed nuclear magnetic resonance technique. Moisture and Frost-Related Properties. Transportation Res. Record 675, p. 11-14.

261. Tice, A., D. Anderson, and K. Sterrett. 1981. Unfrozen water contents of submarine permafrost determined by nuclear magnetic resonance. Eng. Geo. 18:135-146.

262. Matzkanin, G., and C. Gardner. 1974. Nuclear magnetic resonance sensors for moisture measurement in roadways. Frost, Moisture, and Erosion. Transportation Res. Board, Transportation Res. Record 532, p. 77-86.

263. Rollwitz, W. 1965. Nuclear magnetic resonance as a technique for measuring moisture in liquids and solids. A. Wexler (Ed.), Humidity and Moisture. Reinhold Publishing Co., N.Y., NY, 4:149-162.

264. Slichter, C. 1963. Principles of Magnetic Resonance. Harper Row, N.Y.,NY, 397 p.

265. Abragam, A. 1961. The Principles of Nuclear Magnetism. Clarendon Press, Oxford, England, 599 p.

Soil Salinity

266. U.S. Salinity Laboratory Staff. 1954. Diagnosis and improvement of saline and alkali soils. L. Richards (Ed.), U.S. Dept. of Agric. Handbook 60, p. 160.

267. Sonneveld, C., and J. van den Ende. 1971. Soil analysis by means of a 1:2 volume extract. Plant and Soil 35:505-516.

268. Allison, L. 1973. Oversaturation method of preparing saturation extracts for salinity appraisal. Soil Sci. 116:65-69.

269. Beatty, H., and J. Loveday. 1974. Soluble cations and anions. J. Loveday (Ed.), Methods of Analysis of Irrigated Soils. Tech. Communication 54, Commonwealth Bureau of Soils, Commonwealth Agric. Bureau, p. 108-117.

Salinity Sensors

270. Kemper, W. 1959. Estimation of osmotic stress in soil water from the electrical resistance of finely porous ceramic units. Soil Sci. 87:345-349.

271. Richards, L. 1966. A soil salinity sensor of improved design. Soil Sci. Soc. Amer. Proc. 30:333-337.

272. Soilmoisture Equipment Corp. 1980. 5000-A soil salinity sensor. Soilmoisture Equipment Corp., Santa Barbara, CA.

273. Oster, J., and R. Ingvalson. 1967. In situ measurement of soil salinity with a sensor. Soil Sci. Soc. Amer. Proc. 31:572-574.

274. Whitney, M., and T. Means. 1897. An electrical method of determining the salt content of soils. U.S.Dept. of Agric., Bureau of Soils Bull. 8:1-30.

275. Campbell, R., C. Bower, and L. Richards. 1948. Change of electrical conductivity with temperature and the relation of osmotic pressure to electrical conductivity and ion concentration for soil extracts. Soil Sci. Soc. Amer. Proc. 13:66-69.

276. Richards, L., and R. Campbell. 1948. Use of thermistors for measuring the freezing point of solutions and soils. Soil Sci. 65:429-436.

277. Reicosky, D., R. Millington, and D. Peters. 1970. A salt sensor for use in saturated and unsaturated soils. Soil Sci. Soc. Amer. Proc. 34:214-217.

278. Enfield, C., and D. Evans. 1969. Conductivity instrumentation for in situ measurement of soil salinity. Soil Sci. Amer. Proc. 33:787-789.

279. Oster, J., and L. Willardson. 1971. Reliability of salinity sensors for the measurement of soil salinity. Agron. J. 63:695-698.

280. U.S. Salinity Laboratory Staff. 1977. Minimizing salt in return flow by improving irrigation efficiency. U.S. Environmental Protection Agency, EPA-IAG-D5-0370, 128 p.

281. Austin, R., and J. Oster. 1973. Oscillator circuit for automated salinity sensor measurements. Soil Sci. Soc. Amer. Proc. 37:327-329.

282. Ingvalson, R., J. Oster, S. Rawlins, and G. Hoffman. 1970. Measurement of water potential and osmotic potential in soil with a combined thermocouple psychrometer and salinity sensor. Soil Sci. Soc. Amer. Proc. 34:570-574.

283. Wood, J. 1978. Calibration stability and response time for salinity sensors. Soil Sci. Soc. Amer. J. 42:248-250.

284. Wesseling, J., and J. Oster. 1973. Response of salinity sensors to rapidly changing salinity. Soil Sci. Soc. Amer. Proc. 37:553-557.

285. Yadav, B., N. Rao, K. Paliwal, and P. Sarma. 1979. Comparison of different methods for measuring soil salinity under field conditions. Soil Sci. 127:335-339.

Four Electrode Method

286. Halvorson, A., and J. Rhoades. 1974. Assessing soil salinity and identifying potential saline-seep areas with field soil resistance measurements. Soil Sci. Soc. Amer. Proc. 38:576-581.

287. Halvorson, A., and J. Rhoades. 1976. Field mapping soil conductivity to delineate dryland saline seeps with four-electrode technique. Soil Sci. Soc. Amer. J. 40:571-575.

288. Halvorson, A., J. Rhoades, and C. Reule. 1977. Soil salinity four electrode conductivity relationships for soils of the Northern Great Plains. Soil Sci. Soc. Amer. J. 41:966-971.

289. Shea, P., and J. Luthin. 1961. An investigation of the use of the four-electrode probe for measuring soil salinity in situ. Soil Sci. 92:331-339.

290. Roux, P. 1978. Electrical resistivity evaluations at solid waste disposal facilities. U.S. Environmental Protection Agency, SW-729, 93 p.

291. James G. Biddle Co. n.d. Biddle model 63220-Megger Meter. James G. Biddle Co., Plymouth Meeting, PA.

292. Bison Instruments Co. n.d. Bison model 2350B-earth resistivity meter. Bison Instruments Co. Minneapolis, MN.

293. Soil Test, Inc. n.d. Soiltest model R-40-Strata Scout. Soil Test, Inc., Evanston, IL.

294. Austin, R., and J. Rhoades. 1979. A compact, low-cost circuit for reading four-electrode salinity sensors. Soil Sci. Soc. Amer. J. 43:808-810.

295. Moore, R. 1944. An empirical method of interpretation of earth-resistivity measurements. Amer. Inst. of Mining and Metallur. Eng. Tech. Publ. 1743:1-8.

296. Griffiths, D., and R. King. 1956. Applied Geophysics for Engineers and Geologists. Pergamon Press, Elmsford, NY, 223 p.

297. Van Nostrand, R., and K. Cook. 1966. Interpretation of resistivity data. U.S. Geological Survey Prof. Paper No. 499, U.S. Government Printing Office, 310 p.

298. Panofsky, W., 1950, Classical Electricity and Magnetism, U.S. Atomic Energy Commission, UCRL-1014, 347 p.

299. Keller, G., and F. Frischknecht. 1966. Electrical Methods in Geophysical Prospecting. Pergamon Press, Elmsford, NY, 519 p.

300. Rhoades, J., P. Raats, and R. Prather. 1976. Effects of liquid-phase electrical conductivity, water content, and surface conductivity on bulk soil electrical conductivity. Soil Sci. Soc. Amer. J. 40:651-655.

301. Nadler, A. 1981. Field application of the four-electrode technique for determining soil solution conductivity. Soil Sci. Soc. Amer. J. 45:30-34.

302. van Hoorn, J. 1980. The calibration of four-electrode soil conductivity measurements for determining soil salinity. D. Bhumbla and J. Yadav (Eds.), Intl. Symp. on Salt Affected Soils, Feb. 1980, Karnal, India, p. 148-156.

303. Halvorson, A., and A. Reule. 1976. Estimating water salinity with geophysical earth resistivity equipment. Soil Sci. Soc. Amer. J. 40:152-153.

Electrical Conductivity Probes

304. Nadler, A., M. Magaritz, Y. Lapid, and Y. Levy. 1982. A simple system for repeated soil resistance measurements at the same spot. Soil Sci. Soc. Amer. J. 46:661-663.

305. Rhoades, J. 1979. Monitoring soil salinity: A review of methods. L. Everett and K. Schmidt (Eds.), Establishment of Water Quality Monitoring Programs. Proc., Annual Symp., June, 1978, Amer. Water Resources Assn., St. Anthony Falls Hydraulics Lab., Minneapolis, MN, p. 150-165.

306. Rhoades, J., and J. van Schilfgaarde. 1976. An electrical conductivity probe for determining soil salinity. Soil Sci. Soc. Amer. J. 40:647-651.

307. Rhoades, J., and A. Halvorson. 1977. Electrical conductivity methods for detecting and delineating saline seeps and measuring salinity in Northern Great Plains soils. U.S. Dept. of Agric., Agric. Res. Service, ARS W-42, p. 45.

308. Rhoades, J. 1981. Predicting bulk soil electrical conductivity versus saturation paste extract electrical conductivity calibrations from soil properties. Soil Sci. Soc. Amer. J. 45:42-44.

309. Eijkelkamp B.V. Equipment for soil research. Eijkelkamp B.V., Giesbeek, Netherlands.

Inductive Electromagnetic Techniques

310. DeJong, E., A. Ballantyne, D. Cameron, and D. Read. 1979. Measurement of apparent electrical conductivity of soils by an electromagnetic induction probe to aid in salinity surveys. Soil Sci. Soc. Amer. J. 43:810-812.

311. Rhoades, J., and D. Corwin. 1981. Determining soil electrical conductivity-depth relations using an inductive electromagnetic soil conductivity meter. Soil Sci. Soc. Amer. J. 45:255-260.

312. Geonics Limited. 1982. Models EM31 and EM34-3. Geonics Limited, Missisauga, Ontario, Canada.

313. Corwin, D., and J. Rhoades. 1982. An improved technique for determining soil electrical conductivity depth relations from above ground electromagnetic measurements. Draft Paper, U.S. Salinity Lab., U.S. Dept. of Agric., Riverside CA, p. 9.

Temperature

314. The International Practical Temperature Scale of 1968. 1969. Metrologia 5:34-49.

315. The International Practical Temperature Scale of 1968, Amended Edition of 1975. 1976. Metrologia 12:7-17.

316. Stevens, H., H. Ficke, and J. Smoot. 1975. Techniques of water-resources investigations of the United States Geological Survey. Water Temperature Influential Factors, Field Measurement, and Data Presentation, Book I, Chapter D1, p. 63.

317. American Society for Testing and Materials (ASTM). 1981. Manual on the use of thermocouples in temperature measurement. ASTM Special Tech. Publ. 470B, ASTM, Philadelphia, PA, p. 62-80.

318. Tanner, C. 1958. Soil thermometer giving the average temperature of several locations in a single reading. Agron. J. 50:384-387.

319. Oke, T., and F. Hannell. 1966. Variations of temperature within a soil. Weather 21:21-27.

320. Hannell, F., and C. Hannell. 1979. A thermocouple rod designed for the measurement of subsurface temperatures. J. Physics E. Scientific Instruments 12:958-970.

321. Suomi, V. 1957. Soil temperature integrators. H. Lettau and B. Davidson (Eds.), Exploring the Atmosphere's First Mile. Pergamon Press, Elmsford, NY, p. 24.

322. Sargeant, D. 1965. Note on the use of junction diodes as temperature sensors. J. Appl. Meteor. 4:644-646.

323. Becker, J., C. Green, and G. Pearson. 1946. Properties and uses of thermistors—Thermally sensitive resistors. Trans. Amer. Inst. Elec. Engineers 65:1-15.

324. Richards, L., and R. Campbell. 1948. Use of thermistors for measuring the freezing point of solutions and soils. Soil Sci. 65:429-436.

325. Barton, L. 1962. Measuring temperature with diodes and transistors. Electronics 35:38-40.

326. Phene, C., R. Austin, G. Hoffman, and S. Rawlins. 1969. Measuring temperatures with P-N junction diodes. Agric. Eng. 50:684-685.

327. Enfield, C., and J. Tromble. 1970. P-N junctions—A tool for temperature measurement. Water Resources Res. 6:981-985.

328. MacDowall, J. 1957. Soil thermometers. Nature 179:328.

329. Garvitch, Z., and M. Probine. 1956. Soil thermometers. Nature 177:1245-1246.

330. Friedberg, S. 1955. Semiconductors as thermometers. Temperature, Its Measurement and Control in Science and Industry. Reinhold Publishing Corp., N.Y., NY, 2:359-382.

331. Stallman, R. 1967. Flow in the zone of aeration. V. Chow (Ed.), Advances in Hydroscience. Academic Press, N.Y., NY, 4:178-179.

Vacuum Pressure Lysimeters

332. Shuford, J., D. Fritton, and D. Baker. 1977. Nitrate-nitrogen and chloride movement through undisturbed field soil. J. Environ. Qual. 6:255-259.

333. Tyler, D., and G. Thomas. 1977. Lysimeter measurements of nitrate and chloride losses from soil under conventional and no-tillage corn. J. Environ. Qual. 6:63-66.

334. Severson, R., and D. Grigal. 1976. Soil solution concentrations: Effect of extraction time using porous ceramic cups under constant tension. Water Resources Bull. 12:1161-1170.

335. Warrick, W., and A. Amoozegar-Ford. 1977. Soil water regimes near porous cup samplers. Water Resources Res. 13:203-207.

336. Barbarick, K., B. Sabey, and A. Klute. 1979. Comparison of various methods of sampling soil water for determining ionic salts, sodium, and calcium content in soil columns. Soil Sci. Soc. Amer. J. 43:1053-1055.

337. van der Ploeg, R., and F. Beese. 1977. Model calculations for the extraction of soil water by ceramic cups and plates. Soil Sci. Soc. Amer. J. 41:466-470.

338. England, C. 1974. Comments on "A technique using porous cups for water sampling at any depth in the unsaturated zone" by Warren W. Wood. Water Resources Res. 10:1049.

339. Wagner, G. 1962. Use of porous ceramic cups to sample soil water within the soil profile. Soil Sci. 94:379-386.

340. Reeve, R., and E. Doering. 1965. Sampling the soil solution for salinity appraisal. Soil Sci. 99:339-344.

341. Parizek, R. 1970. Soil-water sampling using pan and deep pressure-vacuum lysimeters. J. Hydrology 11:1-21.

342. Apgar, M., and D. Langmuir. 1971. Groundwater pollution potential of a landfill above the water table. Ground Water 9:76-96.

343. Bell R. 1974. Porous ceramic soil moisture samplers, an application in lysimeter studies on effluent spray irrigation. N. Zealand J. Experim. Agric. 2:173-175.

344. David, M., and R. Struchtemeyer. 1980. The effects of spraying sewage effluent on forested land at Sugarloaf Mountain, Maine. Life Sciences and Agric. Experim. Sta. 773:1-16.

345. Yu, Y., K. Chen, R. Morrison, and J. Mang. 1978. Physical and chemical characterization of dredged material sediments and leachates in confined land disposal areas. U. S. Army Corps of Engineers, Tech. Rept. D-78-43, 241 p.

346. Knighton, M., and D. Streblow. 1981. A more versatile soil water sampler. Soil Sci. Soc. Amer. J. 45:158-159.

347. Wood, W. 1973. A technique using porous cups for water sampling at any depth in the unsaturated zone. Water Resources Res. 9:486-488.

348. Wood, W., and D. Signor. 1975. Geochemical factors affecting artificial recharge in the unsaturated zone. Trans. Amer. Soc. of Agric. Engineers 18:677-683.

349. Chow, T. 1977. A porous cup soil-water sampler with volume control. Soil Sci. 124:173-176.

350. Wood, A., J. Wilson, R. Cosby, A. Hornsby, and L. Baskin. 1981. Apparatus and procedures for sampling soil profiles for volatile organic compounds. Soil Sci. Soc. Amer. J. 45:442-444.

351. Grover, B., and R. Lamborn. 1970. Preparation of porous ceramic cups to be used for extraction of soil water having low solute concentrations. Soil Sci. Soc. Amer. Proc. 34:706-707.

352. Aulenbach, D. and N. Clesceri. 1980. Monitoring for land application of wastewater. Water, Air, and Soil Pollution 14:81-94.

353. Wolff, R. 1967. Weathering of woodstock granite near Baltimore, Maryland. Amer. J. Sci. 265:106-117.

354. Hansen, E., and R. Harris. 1975. Validity of soil-water samples collected with porous ceramic cups. Soil Sci. Soc. Amer. Proc. 39:528-536.

355. Stearns, R., T. Tsai, and R. Morrison. 1980. Validity of the porous cup vacuum/suction lysimeter as a sampling tool for vadose waters. Univ. of Southern Calif. Environ. Eng. Lab., CE 513, p. 11.

356. Silkworth, D., and D. Grigal. 1981. Field comparison of soil solution samplers. Soil Sci. Soc. Amer. J. 45:440-442.

357. Dazzo, F., and D. Rothwell. 1974. Evaluation of porcelain cup soil water samplers for bacteriological sampling. Appl. Microb. 27:1172-1174.

358. Quin, B., and L. Forsythe, L. 1976. All-plastic suction lysimeters for the rapid sampling of percolating soil water. N. Zealand J. Sci. 19:145-148.

359. Nielsen, D., and R. Phillips. 1958. Small fritted glass bead plates for determination of moisture retention. Soil Sci. Soc. Amer. Proc. 22:574-575.

360. Chow, T. 1977. Fritted glass bead materials as tensiometers and tension plates. Soil Sci. Soc. Amer. J. 41:19-22.

361. Morrison, R. 1982. A modified vacuum pressure lysimeter for soil water sampling. Soil Sci. 134:206-210.

362. Zimmermann, C., M. Price, and J. Montgomery. 1978. A comparison of ceramic and Teflon® in situ samplers for nutrient pore water determinations. Estuarine Coastal Mar. Sci. 7:93-97.

Vacuum Plates and Tubes

363. Duke, H., E. Kruse, and G. Hutchinson. 1970. An automatic vacuum lysimeter for monitoring percolating rates. U. S. Dept. of Agric., Agric. Res. Serv. ARS 41-165.

364. Tanner, C., S. Bourget, and W. Holmes. 1954. Moisture tension plates constructed from alundum filter discs. Soil Sci. Soc. Amer. Proc. 18:222-223.

365. Cole, D. 1958. Alundum tension lysimeter. Soil Sci. 85:293-296.

366. Cole, D., S. Gessel, and E. Held. 1961. Tension lysimeter studies of ion and moisture movement in glacial till and coral atoll soils. Soil Sci. Soc. Amer. Proc. 25:321-325.

367. Cole, D., and S. Gessell. 1968. Cedar River Research: A program for studying the pathways, rates, and processes of elemental cycling in a forest ecosystem. College of Forest Res., Univ. Wash., Seattle, WA, Inst. of Forest Prod. Contribution 4, p. 54.

368. Haines, B., J. Waide, and R. Todd. 1982. Soil solution nutrient concentrations sampled with tension and zero tension lysimeters: reports of discrepancies. Soil Sci. Soc. Amer. J. 46:658-661.

369. Cochran, P., G. Marion, and A. Leaf. 1970. Variations in tension lysimeter leachate volumes. Soil Sci. Soc. Amer. Proc. 34:309-311.

370. Iskandar, J., and Y. Nakano. 1978. Soil lysimeters for validating models of wastewater renovation by land application. U. S. Army Cold Regions Res. and Eng. Lab., Special Rept. 78-12, 15 p.

371. Duke, H., and H. Haise. 1973. Vacuum extractors to assess deep percolation losses and chemical constituents of soil water. Soil Sci. Soc. Amer. Proc. 37:963-964.

372. van Schilfgaarde, J. 1977. Minimizing salt in return flow by improving irrigation efficiency. Proc. National Conf. Irrigation Return Flow Quality Management. J. Law, Jr. and G. Skogerboe (Eds.), Colorado State Univ., Fort Collins, CO, p. 81-98.

373. U. S. Salinity Laboratory Staff. 1981. Minimizing salt in return flow through irrigation management. U. S. Environmental Protection Agency, EPA-IAG-D6-0370, 160 p.

374. Jackson, D., F. Brinkley, and E. Bondietti. 1976. Extraction of soil water using cellulose-acetate hollow fibers. Soil Sci. Soc. Amer. J. 40:327-329.

375. Levin, M., and D. Jackson. 1976. A comparison of in situ extractors for sampling soil water. Soil Sci. Soc. Amer. J. 40:535-556.

Membrane Filter Samplers

376. Stevenson, C. 1978. Simple apparatus for monitoring land disposal systems by sampling percolating soil waters. Environ. Sci. and Tech. 12:329-331.

377. Shaffer. K., D. Fritton, and D. Baker. 1979. Drainage water sampling in a wet, dual-pore soil system. J. Environ. Qual. 8:241-246.

378. Wagemann, R., and B. Graham. 1974. Membrane and glass fiber filter contamination in chemical analysis of fresh water. Water Res. 8:407-412.

Absorbent Methods

379. Tadros, V., and J. McGarity. 1976. A method for collecting soil percolate and soil solution in the field. Plant and Soil 44:655-667.

380. Shimshi, D. 1966. Use of ceramic points for the sampling of soil solution. Soil Sci. 101:98-103.

PART II
MONITORING IN THE ZONE OF SATURATION

A zone of saturation can exist as a limited saturated area perched at the interface of two soils possessing dissimilar permeabilities or as a ground water system. Monitoring techniques presented for this zone include those in situ devices that retrieve water samples and provide instrument access. Approaches developed for this purpose include:

- drainage systems,
- trench and caisson lysimeters,
- single screened monitoring wells,
- well points,
- single wells and multiple sampling points,
- gas lift samplers,
- hybrid systems, and
- piezometers.

A. DRAINAGE SYSTEMS

Perched or near to surface ground water can be collected by tile or perforated PVC drains. A thin layer of highly permeable gravel or similar medium surrounds most drain designs to induce flow into the collector.[1] In such cases, samples or measurements are obtainable from the outflow or at points along the flow path.

A variety of materials and designs have been employed in drainage systems. The simplest configuration is an irrigation tile drain which collects infiltrating water and transports the water via gravity flow. A similar approach intercepts ground water with a half perforated PVC pipe placed within a gravel filled trench.[2]

Sampling from a drainage system is convenient if such a system exists at a site. Installing drains for the express purpose of sample collection, however, is not recommended due to extensive soil disturbance and the availability of more effective methods.

B. TRENCH AND CAISSON LYSIMETERS

Trench and caisson lysimeters collect water samples from the vadose zone which intermittently becomes saturated due to rainfall, flooding, or irrigation. Most of these devices acquire water by gravity rather than suction.

Trench lysimeter designs belong to either the pan or trough (e.g., Ebermayer) variety. One pan lysimeter design consists of a 30.5 by 38.1 cm metal pan which is flared along one side.[3] Copper tubing is terminated and soldered to the raised edge so that perched water accumulating on the pan drains into the tube. The entire pan is inserted into the side wall of a trench with only the copper tubing protruding. Plastic tubing is connected to the copper through which water drains into a collection vessel.

Ebermayer designs (also called zero tension lysimeters) rely upon a trough or pail rather than a pan.[4] One configuration, shown in Figure 2.1, is a trough upon which a fiberglass screen is suspended so that it hangs in a concave position over the trough.[5] The screen is lined with glass wool and covered with soil. Two parallel bars in contact with the screen bottom are bent towards the collection tube. Water entering the trough is pulled along the bars via capillary forces to the collection tube. The unit is installed in the side wall of a trench.

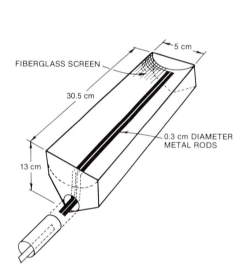

Figure 2.1 Trench lysimeter.[5] (reproduced from SOIL SCIENCE, Volume 105, 1968, page 83,©1968, Williams and Wilkens Co.)

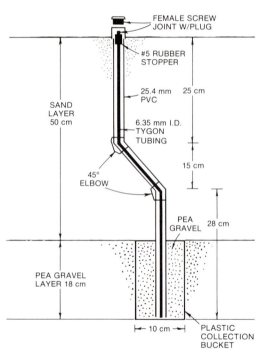

Figure 2.2 Microlysimeter.[7] (reproduced from AGRONOMY JOURNAL, Volume 72, 1980, page 397, by permission of the American Society of Agronomy)

A modification to this design is a metal trough in which a length of perforated PVC pipe is mounted.[6] Graded gravel fills the trough immediately surrounding the pipe with fine sand at the edges and top of the trough. One end of the pipe is capped while the other is connected to a tube for sample drainage. The unit is installed in the side wall of a trench and the sample drains into a collection bottle. A similar device, a microlysimeter (Figure 2.2), relies upon suction for sample retrieval[7] and consists of a small diameter plastic tube slotted at one end. The slotted end of the tube is placed in a plastic bucket filled with pea gravel. The sampling tube is protected by a PVC casing which bends to prevent interference with infiltration directly above the collection bucket. The microlysimeter is primarily usable as part of a prepared soil layer which is sampled.

Caisson lysimeters consist of a vertical chamber from which collectors are placed into the surrounding soil.[8] Figure 2.3 illustrates a design in which a nearly horizontal half screened PVC casing is used to intercept percolating waters.[9] A similar unit is presented in Figure 2.4 in which the caisson is covered by a concrete manhole. Horizontal collection pipes are installed which branch out from the caisson. Samples accumulate in a refrigerated collector until a sufficient volume is available for analysis.[10]

Large sample volumes can be acquired with trench and caisson lysimeters, and the approximate time of collection can be determined. Problems include soil disturbance during installation, alteration of flow paths during construction, and a limited number of large pans that can be used to collect representative samples.[11]

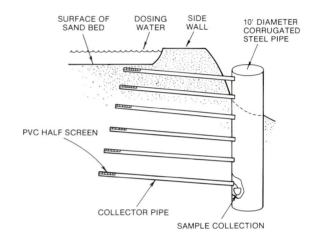

Figure 2.3 Caisson lysimeter.[8]

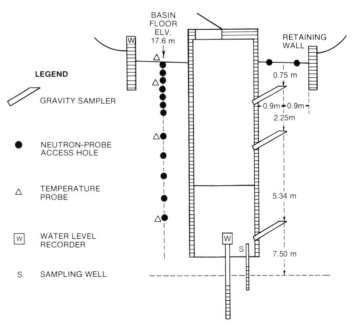

Figure 2.4 Caisson lysimeter with horizontal collector pipes.[9] (reproduced from AIR, WATER, AND SOIL POLLUTION, Volume 14, 1980, by permission of the D. Reidel Publishing Company, Hague, Netherlands)

C. SINGLE SCREENED MONITORING WELLS

A monitoring well provides both access for sample collection and downhole instrumentation. Components of a monitoring well generally include a bottom well plug or point, a length of screen slotted to a particular width and spacing density, blank casing, a well cap, and a protective cover. Figure 2.5 illustrates an example of a single screened monitoring well.

A variety of well designs have been proposed for single screened monitoring wells. Differences arise primarily in the screen length, the choice of seal materials, and the sealing sequence. Ideally, the well is screened the entire depth to where the well intersects the saturated zone, but this is often economically unfeasible. Bentonite or grout is usually selected for creation of an impermeable boundary although various admixtures have been proposed. When using bentonite, a powder or slurry form should be used rather than pellets due to their nonuniform flow and expansion tendencies in the borehole.[12] In order to reduce shrinkage with a grout seal, the cement to water ratio should be minimized.

Several sealing configurations have been proposed and usually consist of a seal the entire length above the well screen or a series of seals interposed with compacted soil. Backfilling entirely with an impermeable seal appears to promote water channeling between the casing and seal interface. Grout is especially susceptible to this potential problem due to temperature changes during the curing process, swelling and shrinkage while the mixture cures, and from poor bonding between the grout and casing surface.[13-15] Similar difficulties may be encountered with bentonite or admixtures. A 15% by volume powdered bentonite with silica sand appears to minimize some of these difficulties. Irrespective of the seal sequence, the material should be pumped into the annulus to insure proper material distribution.

A surface seal should be mounded around the well casing as settlement of the backfilled hole generally occurs. A plastic or steel security cover is placed over the excavated hole and should be anchored with a cement base (Figure 2.6). The cover protects the well casing and reduces surface water drainage into the backfilled hole. Since the disturbed soil immediately adjacent to the borehole annulus is often more permeable than the indigenous, undisturbed soil (e.g., from drilling), surface waters tend to percolate along this pathway. A properly installed security cover can minimize surface infiltration near the well head by enclosing the soil adjacent to the borehole.

Excavation may be accomplished by a variety of drilling approaches suited to a particular hydrogeologic setting. These approaches, along with development methods for cleaning the side wall and soil formation in the vacinity of the screen, are discussed in the literature.[16-18]

Materials used in well construction can bias the water quality of a sample; selection is therefore predicated upon the parameters to be analyzed. Common materials

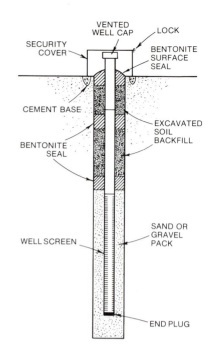

Figure 2.5 Single screened monitoring well. (illustration © 1983, TIMCO MFG., INC., Prairie du Sac, WI)

Figure 2.6 Security cover. (photograph © 1983, TIMCO MFG., INC., Prairie du Sac, WI)

include a high grade stainless steel, PVC, ABS (acrylo-nitrile butadiene styrene), CAB (cellulose acetate butyrate), and Teflon® (tetrafluoroethylene polymer). Of these materials, stainless steel and PVC are the most common.

Stainless steel is often used in applications when specific major ions are of interest. The use of stainless steel as screen and casing material for organics has been proposed: its use for trace metals in the form of a microbiological nutrient or inhibitor may also affect the microflora population within the well.[19]

PVC is the most widely used material for monitoring well construction and probably causes fewer monitoring interferences with volatile organics than other plastic casings. The degree of absorption and leaching appears to be specie (chemical) related. Nevertheless, degradation of PVC with the resultant increase in crystallinity, decreased tensile strength, and the potential release of napthalene, butyloctylfumarate, butylchloroacetate, etc., will eventually occur.[20,21] The degree to which the catabolization of the carbon source within the PVC and resulting CO_2 liberation affects the water within the well is expected to be minimal due to the low biodegradation rates associated with this process.

An ideal material for well construction is Teflon®.[22] Teflon® granular resins used for well casing or screens are often either tetrafluoroethylene (TFE) polymers or tetrafluoroethylene hexafluoropropylene (FEP-fluorocarbons) copolymers. Due to the high cost of Teflon®, many wells are constructed with a Teflon® well screen threaded into PVC casing above the water table level. Such a well provides the optimum combination of materials for a monitoring well.*

Well casing or screen lengths may be joined by solvent welding of the sections with a slip coupling, a bell and spigot arrangement, and threaded adaptors which are solvent welded to the casing or screen section. Irrespective of the joining technique, a uniform inner and outer casing diameter should be maintained. Inconsistent inner diameters result in problems when downhole samplers or instruments with little clearance are lowered into the well. An uneven outer diameter tends to promote water migration at the casing and seal interface to a greater degree than is experienced with uniform outer diameter casing. Other problems associated with these joints include inconsistent bonding between the two sections and the potential contribution of volatile organic species from the solvent [e.g., N, N-dimethylformamide (DMF), methyl-isobutylketone (MIBK), methylethylketone (MEK), tetrahy-drofuran (THF), and cyclohexane]. [23-25] Since the type of solvent or glue used for bonding varies with the type of casing material, the chemical composition of the material should be examined in relation to the immediate and possible long term water quality monitoring goals. If the material is suspected of directly contributing to or interacting with water quality analysis, a sample of each bonding agent should be tested by mechanical stirring with unchlorinated water equal to 10 times its volume for 24 hours, followed by gas chromatograph/mass spectrometer (GC/MS) analysis[26] of the water extract.[27]

Publisher's note:

The practice of over simplification of a monitoring well must be corrected and its value as an investigative tool be realized.

PVC materials offer extreme cost advantages while giving excellent results under given situations in lowered concentrations of organic materials and trace metals. However, we have experienced poor results in applications where high concentrations of benzenes, chlorinated solvents and other volatile organics were present. The importance of associating proper chemical resistant materials must be emphasized when designing monitoring programs suitable for use in both organic and inorganic situations. Due to its high chemical resistance TFE resin, a polymer consisting of recurring tetrafluoroethylene monomer units (CF_2-CF_2)m, is an excellent choice.

Several methods of using Teflon® tubes and slotted screens are suggested which will allow for maximum cost efficiency whenever possible.

Some typical monitoring well material combinations are: a PVC Monitoring Well with PVC riser tube, screen, and plug; a PVC and Teflon® Monitoring Well with PVC riser tube, Teflon® riser tube, Teflon® screen and plug; and a Stainless Steel and PVC Monitoring Well with PVC riser tube, stainless steel riser tube, stainless steel screen and plug.

We have concluded that in order to have monitoring wells which are expected to give high quality analytical data, we must eliminate shortcuts and materials of mediocre workmanship. It seems foolhardy to save pennies at the expense of quality.

It is ludicrous at best to install sample wells constructed with possible contaminate contributing materials then spend thousands of dollars for high level analytical work, sample transportation, etc. and then have the audacity to state that the results are representative of the area of investigation.

The installation of a monitoring well system must be done with the knowledge and understanding of the investigative purposes of the well.

Excerption from R. Timmons. 1983. Monitoring Wells, An Investigative Tool. ©1983, Timco Mfg., Inc., Prairie du Sac, WI 53578. Presented at the First National Symposium and Exposition on Ground-Water Instrumentation in Las Vegas, Nevada, on March 16, 1983.

Casing with threads machined directly into the pipe eliminates potential problems associated with solvent welding by providing solventless construction and a flush joint between the inner and outer pipe diameters. While various threads, (e.g., Acme, buttress, or standard pipe thread) have been employed, the square flush jointed thread illustrated in Figure 2.7 is best suited for monitoring well construction. When two flush threaded pipes are joined, the inner and outer diameters become uniform along the pipe length.

The tensile strength of the joint is important as it will determine the maximum axial load that can be placed along the casing axis for a particular material.[28-29] Tensile strengths of flush jointed, 8 non-modified square threads per inch are listed in Table 2.1. By dividing the tensile strength by the per linear weight of PVC, the maximum theoretical depth to which a dry string of casing can be suspended in a hole is calculable. In most cases, the casing will encounter water. This increases the length of the casing that can be suspended by approximately 40% (the specific gravity of PVC = 1.4) for that portion of the casing in contact with the water.

Figure 2.7 Flush jointed square threads. (photograph ©1983, TIMCO MFG.,INC., Prairie du Sac, WI)

TABLE 2.1   Tensile Test Results of PVC Casing with 8 Non-Modified Square Flush Threads

Inside Diameter (in)	Nominal Weight Per 100 ft.	Ultimate Load (lb)	Suggested Maximum Dry String Depth* (ft)
½ Schedule 80	20.5	635	2,787.8
1¼ Schedule 80	56.7	1,740	2,761.9
1½ Schedule 40	51.8	2,485	4,317.5
2 Schedule 40	69.5	2,650	3,431.6
2 Schedule 80	94.9	2,895	2,745.5
3 Schedule 40	143.5	5,600	3,512.2
4 Schedule 40	204.3	8,900	3,920.7
4 Schedule 80	283.3	11,200	3,558.0
5 Schedule 40	277.6	10,500	3,404.2
5 Schedule 80	393.8	12,500	2,856.7
5 Class 200	187.6	11,000	5,277.2
6 Schedule 40	360.0	11,050	2,762.5
6 Schedule 80	541.1	15,650	2,603.0

*A 10% safety factor has been assumed in these suggested lengths.

Table 2.1 (reproduced from TIMCO GEOTECHNICAL PRODUCTS CATALOGUE, 1982, by permission of Timco Mfg., Inc., Prairie du Sac, WI)

While a single screened well is the traditional ground water approach for monitoring, a number of inherent limitations exist. One of the most important is the inability to determine the precise entry point of a contaminant into the well. The vertical distribution of the contaminant within the water column, therefore, cannot be precisely determined as contaminant stratification may be masked.

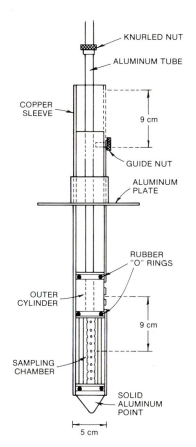

KNURLED NUT

ALUMINUM TUBE

COPPER SLEEVE

9 cm

GUIDE NUT

ALUMINUM PLATE

RUBBER "O" RINGS

OUTER CYLINDER

9 cm

SAMPLING CHAMBER

SOLID ALUMINUM POINT

5 cm

Figure 2.8 Well point sampler.[32] (reproduced from R. Summerfield, PLANT AND SOIL, Volume 38, 1973, by permission of Martinus Nijhoff Publishers B.V.)

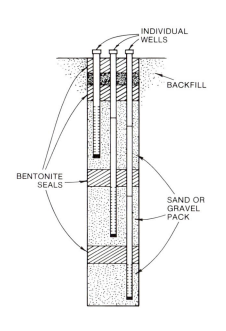

INDIVIDUAL WELLS

BACKFILL

BENTONITE SEALS

SAND OR GRAVEL PACK

Figure 2.9 Well cluster.

D. WELL POINTS

A well point is usually fabricated from metal and includes a screen, casing, and hardened point. The entire assemblage is jetted or driven into the soil; when the proper depth is reached, the well point is left in place to function as a monitoring well.

Numerous modifications to the basic well point have been developed for monitoring purposes.[30,31] An example of one method, shown in Figure 2.8,[32] has the probe pushed into the soil until the aluminim plate rests on the surface. Water entering the sample chamber is collected by suction through an aluminum tube. The well point can be forced to a deeper level and the procedure repeated. A similar method has been used with the exception that a small bailer is lowered within the access tubing.[33] Samples have been recovered from 27 m with this device.

A sampler with a ceramic tube attached above the well point has been used for sampling dissolved gases in the ground water.[34] The well points installed to the proper depth and filled with water of a known pH. The water within the cup is allowed to reach equilibrium with the ground water via diffusion through the ceramic. Once the pH of the ground water and the sample are identical, the water is removed.

Well point methods offer a portable, quick, and efficient method for monitoring at shallow depths in bogs, muds, unconsolidated sands, and permafrost. The technique has certain depth limitations and cannot be used in consolidated soils. Forcing the unit into the soil also affects the soil density immediately adjacent to the well which may influence some downhole monitoring instruments.

E. WELL CLUSTERS

A well cluster is composed of several small diameter wells which terminate at different depths within a single borehole. A more accurate portrayal of vertical contaminant stratification is therefore possible than with a single screened well.

Figure 2.9 depicts a typical well cluster used to sample at discrete depths. Installation techniques and appropriate materials are identical to the procedures suggested for a single screened monitoring well. Special care must be exercised to isolate the various screen sections to ensure the representativeness of each well sample at a particular depth.

Well clusters provide a means of profiling the vertical distribution of a contaminant in one borehole although the difficulty of successfully isolating the screened portions is a problem. The increased surface area from several small diameter wells also increases the potential channeling of surface water or water between the wells along the casing and soil or seal interface.

73

F. SINGLE WELLS WITH MULTIPLE SAMPLING POINTS

There are four common approaches using a single well to obtain water samples at various depths (multiple sampling points): wells with perforations or screens at various intervals; a well in which sampling probes are placed; a well casing with individual sampling ports; and a bundle type well.

A well may be screened or perforated at predetermined intervals to allow ground water entry at these depths. Screens can be positioned between lengths of casing prior to installation or blank casing can be placed into the borehole and perforated with a variety of downhole devices. A bentonite or grout seal is placed between the screens.

Well casing can also be used to position sampling probes within the borehole. In one design, a screened well point with multiple sampling probes within the casing is used for shallow (< 9.1 m) sampling or piezometric measurements (Figure 2.10). The unit consists of a series of fiberglass probes (Figure 2.11) placed within a 3.2 cm diameter well point. Samples are withdrawn by attaching each sample tube to a two way stoppered container. A common vacuum line is connected to each sample container for sample withdrawal. Another approach withdraws the casing while the probes remain. The probes consist of an Alundum thimble sealed with a rubber stopper through which an extraction tube is attached.[37]

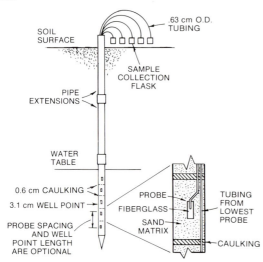

Figure 2.10 Ground water profile sampler.[35] (reproduced from E. Hansen and A. Harris, WATER RESOURCES RESEARCH, Volume 10, page 375, 1974, copyrighted by the American Geophysical Union)

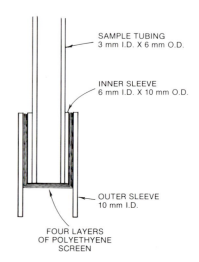

Figure 2.11 Polyethylene covered sampling probe.[36] (reproduced from E. Hansen and A. Harris, WATER RESOURCES RESEARCH, Volume 16, 1980, copyrighted by the American Geophysical Union)

A third approach uses a well casing with individual sampling points. One design has 2.1 cm diameter holes drilled at various intervals in a length of PVC pipe.[38] Each hole is glued with a one hole rubber stopper from within the casing; a screen or fiberglass and cloth layer covers the exterior opening. Polypropylene tubing connects the stopper to the ground surface (Figure 2.12). Bentonite seals are used to isolate each sampling port.[40-42] An initial volume equal to that in the sampling tube is withdrawn and discarded prior to sample collection. For sampling depths greater than 9.1 m, one adaptation that relies on negative and positive gas pressure is used for sample retrieval.[43] Tubing attached to the sampling port is connected to a check valve that is fitted to a polyethylene "T" connector. One T outlet joins a second check valve. The other attaches to a 50 cm plastic syringe (Figure 2.13). To operate, a vacuum is applied which causes valve B in Figure 2.13 to close as valve A opens and draws water into the syringe. Positive pressure forces the syringe plunger down, closes valve A, and opens valve B, thereby forcing the sample to the ground surface. The design is capable of sampling to depths of 50 m.

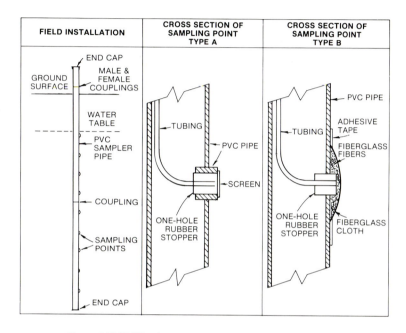

Figure 2.12 Multilevel ground water sampler.[39] (© 1979, reproduced from GROUND WATER, Volume 17)

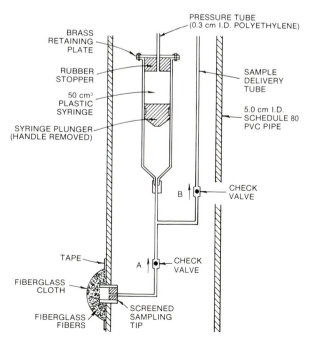

Figure 2.13 Positive displacement sampling point.[43]

The fourth approach is the bundle monitoring well (Figure 2.14). One such well consists of eight polyethylene tubes arranged around a central PVC tube.[44] The end of the center tube is slotted and covered with 200 mesh nylon; each perforated polyethylene tube is also covered with this nylon. A sample is collected by suction (shallow depths) or with a bailer.

Wells screened or perforated at various intervals require special packers or similar equipment to isolate each sampling section. Whether the water sample is representative of that portion of the water column opposite the open area is suspect as water can be drawn from above or below the casing opening.[45] The metal well point with probe approach (Figure 2.10) allows the entire unit to be driven, which may result in reduced soil disturbance. Use of the metal well point, however, precludes its suitability for analysis of some trace metals such as iron and zinc.

The multilevel sampler with external sampling ports is more suitable for most heavy metal analysis, but the sampler cannot be driven into the soil during installation. The number of sampling ports is limited by the number of polyethylene tubes that can be inserted into the casing.

G. GAS LIFT SAMPLERS

A gas lift sampler is installed in situ and relies upon gas pressure to lift the sample to the surface. While a variety of designs exist, check valves that allow water entry but close upon injection of a gas into the vessel are a common feature.[46-49] For shallow (<9.1 m) monitoring applications, samples from units without a check valve arrangement can be collected with suction.[30,50]

The basic components of a gas lift sampler are shown in Figure 2.15. All the sections are threaded to facilitate pre-installation cleaning; a unit can therefore be assembled in the field so that chamber size, screen density and width, and overall geometry can be tailored to a particular need (Figure 2.16). The cap is fitted with a threaded eye hook connected to a suspension line for lowering the sampler into the well or threads machined into the cap (Figure 2.15b). The latter design allows direct threading of the sampler onto a section of casing.

A gas lift sampler can be installed individually or in a multiple sampling arrangement. For a single installation, a hole is excavated to the desired depth and about 30 cm of gravel or sand is poured into the borehole. The sampler is centered above this layer by either the suspension line, a long rod threaded into the eye hook hole, or by utilizing the threaded cap design. A minimum distance of 5 cm should be left between the sampler and borehole sidewall. This annulus is backfilled to at least 30 cm above the unit upon which an impermeable seal is placed. The pressure and extraction tubes are collected and labeled accordingly at the surface. For multiple installations within a single borehole, this procedure is repeated.

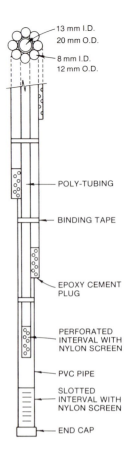

13 mm I.D.
20 mm O.D.

8 mm I.D.
12 mm O.D.

POLY-TUBING

BINDING TAPE

EPOXY CEMENT PLUG

PERFORATED INTERVAL WITH NYLON SCREEN

PVC PIPE

SLOTTED INTERVAL WITH NYLON SCREEN

END CAP

Figure 2.14 Bundle monitoring well.[44]

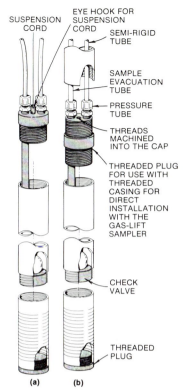

Figure 2.15 Gas lift sampler. (reproduced from TIMCO GEOTECHNICAL PRODUCTS CATALOGUE, 1982, by permission of Timco Mfg., Inc., Prairie du Sac, WI)

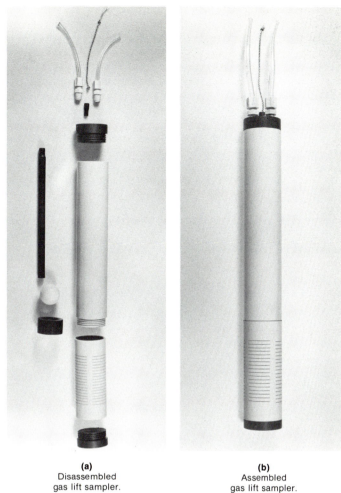

(a)
Disassembled
gas lift sampler.

(b)
Assembled
gas lift sampler.

Figure 2.16 Disassembled and assembled gas lift sampler. (photographs © 1983, TIMCO MFG., INC., Prairie du Sac, WI)

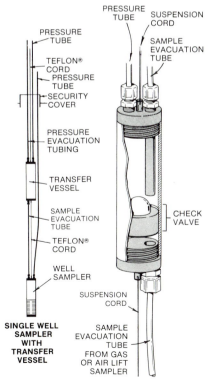

Figure 2.17 Transfer vessel. (reproduced from TIMCO GEOTECHNICAL PRODUCTS CATALOGUE, 1982, by permission of Timco Mfg., Inc., Prairie du Sac, WI)

The design in Figure 2.16 has collected ground water samples as deep as 47.5 m. To achieve greater sampling depths, a transfer vessel (Figure 2.17) is attached to the gas lift sampler to provide a staged sample retrieval system. This arrangement can be expanded for the placement of multiple sampling units at depth within a single borehole. In addition to greater sampling depths, the pressure required for sample transport is reduced.

Prior to sample collection, any water within the gas lift sampler is cleared by injecting gas into the pressure tube. In the case of a gas lift sampler and transfer vessel combination, water from the sampler enters the transfer vessel via the evacuation tube of the gas lift sampler. The transfer vessel is pressurized and the water transported to the surface. During pressurization of the transfer vessel, the gas lift sampler is recharged with water. The time required for vessel recharge depends upon the depth of the sampler below the piezometric surface, the screen slot size and density, and vessel size.

The advantages of a gas lift sampler include an all PVC and Teflon® solventless construction, minimum oxidation during sample transport, and multiple sampling capabilities within a saturated zone. Purging the system prior to sample collection with an inert gas removes most of the oxygen in the system, thereby reducing oxidation reactions. If optimal pressures are used for vessel pressurization, the sample is transported to the surface with minimal sample agitation and mixing except at the interface of the sample and the pressurized gas. Since the time of water entry into the unit can be precisely determined, a representative portrayal of the vertical concentration of a pollutant at a specific point in time can be derived. A device similar to a manometer can be used to obtain piezometric measurements with the sampler.

A problem with gas lift samplers is the entry of fine particles into the check valve and subsequent clogging; water is then forced through the check valve opening rather than through the extraction tube. Proper backfilling and slot selection can minimize this problem.[47] The use of pressure for sample retrieval may also be unsuitable for certain analysis.

H. HYBRID WELL SYSTEMS

Hybrid wells refer to the combination of compatible vadose and saturated zone sampling devices placed within a single borehole. While many combinations are feasible with two separate excavations, the spatial variability and heterogeneous nature of the soil system suggests that, in many cases, samples from a continuous vertical profile are preferred to samples from multiple holes. Three hybrid arrangements are: a small diameter well and lysimeter, a gas lift sampler and lysimeter, and a well and sleeve lysimeter. Figure 2.18 illustrates all three approaches.

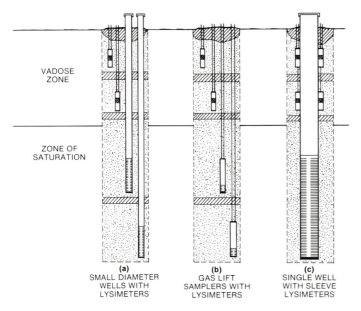

VADOSE
ZONE

ZONE OF
SATURATION

(a)
SMALL DIAMETER
WELLS WITH
LYSIMETERS

(b)
GAS LIFT
SAMPLERS WITH
LYSIMETERS

(c)
SINGLE WELL
WITH SLEEVE
LYSIMETERS

Figure 2.18 Hybrid well systems.

ELECTRICAL
SIGNAL
LINES

FLEXIBLE
CABLE

AMPLIFIER

PNEUMATIC
SIGNAL
LINES

STAINLESS
STEEL
HOUSING

0 OR NULL
CALIBRATION

EPOXY
POTTING
COMPOUND

ELECTRICAL
TRANSDUCER

PNEUMATIC
PIEZOMETER

FILTER STONES

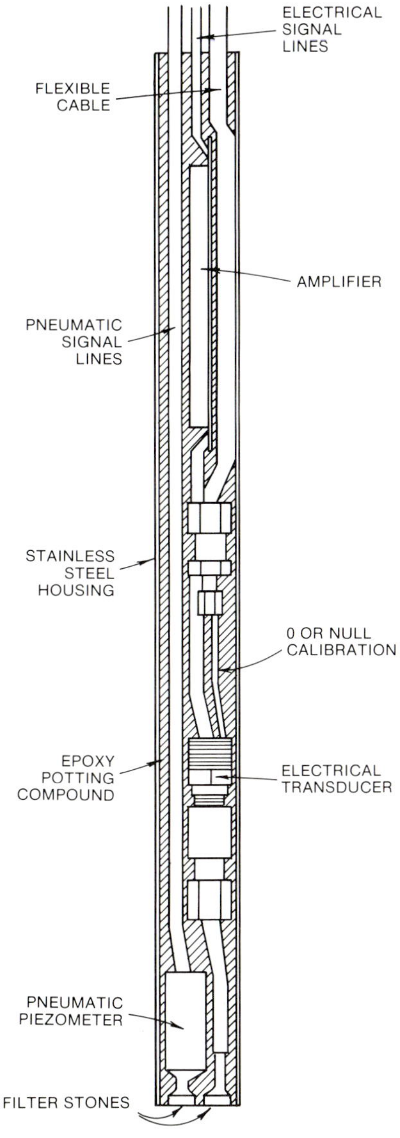

Figure 2.19 Pneumatic piezometer with vibrating wire.[53] (reproduced from PETUR MODEL PNE-100 ELECTRO-PNEUMATIC PIEZOMETER, 1981, by permission of Petur Instrument Co., Inc.)

In the first arrangement, one or more small diameter (3.1 cm) wells are screened within the saturated zone (Figure 2.18a). A single well with multiple sampling points may also be used. A bentonite or grout seal isolates the portion of the well below the water table from the upper portion of the borehole. One or more lysimeters are placed above this seal within the vadose zone; impermeable seals are used to isolate each lysimeter. A surface seal completes the installation.

A series of gas lift samplers positioned in the saturated zone and lysimeters in the vadose zone consitutes another hybrid well system (Figure 2.18b). In this arrangement, sampling is limited to suction or pressurization techniques.

A single well and sleeve lysimeter (Figure 2.18c) combines the lysimeter and well casing without interfering with the inner diameter of the well. The well is screened along the appropriate interval within the saturated zone; sleeve lysimeters are threaded onto the well casing above the screen above the desired interval.

Hybrid well systems maximize the use of a single borehole to provide a vertical profile of a contaminant throughout the vadose and saturated zone. Since the greatest cost of most monitoring installations is the drilling, hybrids are extremely cost effective. The major difficulty in hybrid systems is proper isolation of the various sampling points, especially since numerous individual wells or tubing surfaces exist for water channelization.

I. PIEZOMETERS

A piezometer measures pressure. In some cases, the unit is designed so that both pressure measurements and water samples can be obtained. Piezometers are classified as either pore pressure piezometers or well piezometers.

1. Pore Pressure Piezometers

Three types of pore pressure piezometers include the electric vibrating wire, hydraulic, and pneumatic.[51] A vibrating wire piezometer is connected to a diaphragm situated behind a filter stone. Changes in pore pressures increase or decrease wire tension which is amplified and transferred to the surface for recording. Calibration is required prior to installation for single or multiple installations.[52]

A device which combines a silicon vibrating wire gage with a pneumatic piezometer in one unit is shown in Figure 2.19.[53] Since both parts measure pore pressure, in situ calibration and a function check of the vibrating wire gage occurs.[54]

The two types of hydraulic piezometers have either a single or double tube. A double tube piezometer consists of a porous stone connected to two tubes which terminate at the surface. High or low entry ceramic or porous stone in different geometries are used depending upon soil conditions. The double tube piezometer is considered superior to the single tube since the unit can be flushed with distilled water if the sensor becomes clogged with fine grained sediments. The double tube hydraulic peizometer can also be circulated

79

with deaired water until the tubes and porous tip are filled. The system remains filled with water though it should be flushed occassionally to remove air which has entered the porous tip through diffusion. Pore water pressure at the piezometer tip can be measured from either tube with the associated corrections for head difference between the tip and measuring gage. A mercury manometer, Bourdon gage, or transducer can be used with an automatic scanner and printer for reading a series of units.

A pneumatic piezometer consists of a diaphragm and a check valve. As air is forced down a tube past the check valve to the diaphragm, an equal pressure is exerted on the diaphragm by the water. Water pressure is measured by an electrical pressure transducer which uses a semiconductor pressure sensor.[55] One pneumatic piezometer design is depicted in Figure 2.20. A portable digital indicator or automatic multichannel data acquisition system records the measurements.

Installation procedures for all pore pressure devices are similar. Upon borehole excavation, approximately 30 cm of fine sand is placed at the bottom of the excavation. The sensor is placed above the sand and covered with additional sand to the depth desired for measurement. A bentonite seal isolates the unit within the borehole from other units.

Pore pressure piezometers operate well with an electronic data acquisition system. Such an array can provide immediate hydraulic head measurements for a large area. Multiple probes can be installed within a borehole or in locations unavailable to a well piezometer.

Problems associated with pore pressure piezometers include clogging of the tubes or corrosion of the transducer in the case of pneumatic designs. Calibration of electronic and pneumatic piezometers by the manufacturer can result in difficulties when additional cable or tubing is required.

2. Well Piezometers

A well piezometer (also referred to as a standpipe piezometer) usually consists of a small diameter PVC or metal casing which is either open at its terminus or connected to a screened well point. Piezometer tip materials include galvanized steel, bronze, porous plastic, ceramic, and Casagrande stone. Water level measurements can be obtained from the well piezometer with chalked tape, acoustic methods, air lines, electric probes, resistance elements, Bourdon gages, manometers, or transducers.[57-59]

Well piezometers can be installed by augering, jetting, or cable tool methods; the selection of piezometer casing material is dependent upon the installation technique. An important step in the installation process is sealing the borehole above the screened portion of the well. A bentonite or grout slurry is poured above the screened section with a tremie pipe above which sand is placed[61] (Figure 2.21). The grout seal may also be extended to the surface to ensure seal contact between the casing and soil wall. Figure 2.22 depicts a standpipe piezometer in which a grout seal above the bentonite extends to the surface. Several piezometers

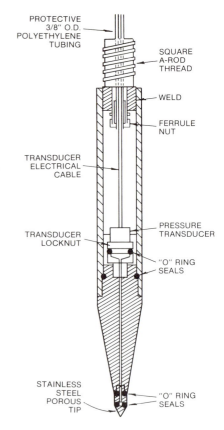

Figure 2.20 Piezometer probe.[56]

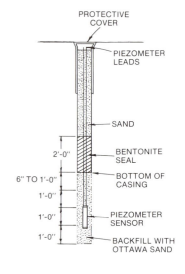

Figure 2.21 Piezometer installation.[60] (reproduced from H. Hemond, WATER RESOURCES RESEARCH, Volume 18, pages 182-186, 1982, copyrighted by the American Geophysical Union)

can also be placed within a single borehole. In this arrangement, the screened portion of each unit is surrounded with sand and the area between the screens is backfilled with bentonite.

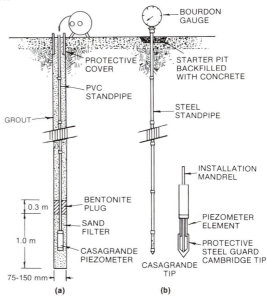

Figure 2.22 (a) Borehole standpipe piezometer and **(b)** drive-in standpipe piezometer.[62] (reproduced from STANDPIPE PIEZOMETER, Data Sheet 1W. 1981, by permission of Solinst Canada, Ltd.)

Well piezometers can provide hydraulic head measurements in a static or nonequilibrium condition, although a time lag occurs when the piezometric level, particularly in an anisotropic soil, is in flux. The time lag is defined as:

$$T = \frac{d^2 \ln\left(\dfrac{mL}{D}\right) + \left(1 + \dfrac{mL}{D}\right)^2}{8 \cdot L \cdot K_h} \qquad (2.1)$$

where T = basic hydrostatic time lag,

d = diameter of pipe above screened section,

L = length of pipe above screened section,

D = diameter of intake screen,

K_h = horizontal permeability, and

$m = \dfrac{K_h}{K_v}$, where K_v = vertical permeability.[63]

Since the pipe diameter (d) is proportional to (T), a small opening significantly increases piezometer sensitivity. Smaller diameter casing or the use of insertable reducers will decrease this basic hydrostatic time lag.[64]

Well piezometers offer a simple and dependable method of obtaining water measurements. These measurements, however, are obtained during one interval compared to other approaches in which a number of sensors are placed within a single small diameter borehole. A well piezometer also does not offer the versatility affordable with pore pressure piezometers.

REFERENCES — PART II

1. Kirkham, D., S. Toksöz, and R. van der Ploeg. 1974. Steady flow to drains and wells. J. van Schilfgaarde, R. Dinauer, M. Davis (Eds.), Drainage for Agriculture. Agronomy, 17 in series monographs. Amer. Soc. of Agron., Madison, WI, p. 203-244.

2. Tolman, A., A. Ballestero, W. Beck, and G. Emrich. 1978. Guidance manual for minimizing pollution from waste disposal sites. U. S. Environmental Protection Agency, EPA-600/2-78-142, 83 p.

Trench and Caisson Lysimeters

3. Parizek, R., and B. Lane. 1970. Soil-water sampling using pan and deep pressure-vacuum lysimeters. J. Hydrology 11:1-21.

4. Kohnke, H., F. Dreibelbis, and J. Davidson. 1940. A survey and discussion of lysimeters and bibliography on their construction and performance. U.S. Dept. of Agric., Misc. Publ. 372, p. 68.

5. Jordan C. 1968. A simple, tension-free lysimeter. Soil Sci. 105:81-86.

6. Moore, B., B. Sagik, and C. Sorber. 1981. Viral transport to ground water at a wastewater land application site. J. Water Pollution Control Fed. 53:1492-1501.

7. Drake, R., I. Pepper, G. Johnson, and W. Kneebone. 1980. Design and testing of a new microlysimeter for leaching studies. Agron. J. 72:397-398.

8. Schmidt, C. and E. Clements. 1978. Reuse of municipal wastewater for groundwater recharge. U. S. Environmental Protection Agency, 68-03-2140, Cincinnati, OH, p. 110-125.

9. Aulenbach, D., and N. Clesceri. 1980. Monitoring for land application of wastewater. Water, Air, and Soil Pollution 14:81-94.

10. Vaughn, J. and E. Landry. 1978. State of knowledge in land treatment, II. Intl. Symp. U.S. Army Corps of Engineers, Cold Regions Res. and Tech. Lab., Hanover, NH, p. 233-243.

11. Krone, R., H. Ludwig, and J. Thomas. 1951. Porous tube device for sampling soil solutions during water-spreading operations. Soil Sci. 73:211-219.

Single Screened Monitoring Wells

12. Fetzer, C. 1982. Pumped bentonite mixtures used to seal piezometers. J. Geotech. Eng. Div., ASCE 108:295-299.

13. Molz, F., and C. Kurt. 1979. Grout-induced temperature rise surrounding wells. Ground Water 17:264-269.

14. Johnson, R., C. Kurt, and G. Dunham, Jr. 1980. Well grouting and casing temperature increases. Ground Water 18:7-13.

15. Kurt, C., and R. Johnson, Jr. 1982. Permeability of grout seals surrounding thermoplastic well casing. Ground Water 20:415-419.

16. Campbell, M., and J. Lehr. 1973. Water Well Technology. McGraw-Hill Book Co., N.Y., NY, 681 p.

17. U.S. Environmental Protection Agency. 1975. Manual of water well construction practices. Office of Water Supply, EPA-570/9-75-001, 156 p.

18. Scalf, M., J. McNabb, W. Dunlap, R. Cosby, and J. Fryberger. 1981. Manual of ground-water quality sampling procedure. Robert S. Kerr Environ. Res. Lab., Office of Res. and Devel., U.S. Environmental Protection Agency, 93 p.

19. Geraghty and Miller, Inc. 1977. The prevalence of subsurface migration of hazardous chemical substances at selected industrial waste disposal sites. U.S. Environmental Protection Agency, EPA-530/SW-634, 166 p.

20. Albertsson, A. 1978. Biodegradation of synthetic polymers, II: A limited microbial conversion of 14C (carbon isotope) in polyethylene to 14 CO_2 (carbon dioxide isotope) by some soil fungi. J. Appl. Polymer Sci. 22:3419-3433.

21. Albertsson, A., Z. Banhidi, and L. Ericcson. 1978. Biodegradation of synthetic polymers, III: The liberation of CO_2 (carbon dioxide isotope) by molds like *Rusarium redolens* from 14C (carbon isotope) labeled pulverized high-density polyethylene. J. Appl. Polymer Sci. 22:3435-3437.

22. DuPont de Nemours and Co., Inc. 1981. TEFLON® fluorocarbon resins. Mechanical Design Data, Fluoropolymers Div., DuPont de Nemours and Co., Inc., Wilmington, DL, 61 p.

23. Reich, K., A. Trussell, F. Lieu, L. Leong, and R. Trussel. 1981. Diffusion of organics from solvent-bonded plastic pipes used for potable water plumbing. American Water Works Conferences, St. Louis, MO, 8:1100-1106.

24. Sosebee, J., P. Geiszler, D. Winegardner, and C. Fischer. 1982. Contamination of groundwater samples with PVC adhesives and PVC primer from monitor wells. Environmental Science and Engineering, Gainesville, FL, DEP 14.6/PVCASTM 24 p.

25. Miller, G. 1982. Uptake and release of lead, chromium, and trace level volatile organics exposed to synthetic well casing. Proc. Second National Symp. on Aquifer Restoration and Ground Water Monitoring, Columbus, OH, p. 236-245.

26. Willard, H., L. Merritt, and J. Dean. 1974. Instrumental Methods of Analysis. D. Van Nostrand Co., N.Y., NY, p. 528-550.

27. Aberdeen Proving Ground. 1981. Tooele Army Depot request for proposal. U.S. Army Toxic and Hazardous Materials Agency, DAAG49-81-R-0044, p. 14(g).

28. Kurt, C., and W. Pate. 1981. Tensile strength of PVC and ABS well casing joints. J. Irrigation and Drainage Div., ASCE 107:197-209.

29. American Society for Testing and Materials. 1982. Annual Book of ASTM Standards, Part 34: Plastic Pipe and Building Products. ASTM, Philadelphia, PA, p. 1110.

Well Points

30. John, P., M. Lock, and M. Gibbs. 1977. Two new methods for obtaining water samples from shallow aquifers and littoral sediments. J. Environ. Qual. 6:322-324.

31. Harrison, W., T. Osterkamp, and M. Inoue. 1981. Details of a probe method for interstitial soil water sampling and hydraulic conductivity and temperature measurement. Univ. of Alaska at Fairbanks, Geophys. Inst. Rept. No. UAG R-280, p. 22.

32. Summerfield, R. 1973. A probe for sampling mire waters for chemical and gas analysis. Plant and Soil 38:469-472.

33. Harrison, W., and T. Osterkamp. 1981. A probe method for soil water sampling and subsurface measurements. Water Resources Res. 17:1731-1736.

34. Rutter, A., and J. Webster. 1962. Probes for sampling ground water for gas analysis. J. Ecology 50:615-618.

Single Wells with Multiple Sampling Points

35. Hansen E., and A. Harris. 1974. A groundwater profile sampler. Water Resources Res. 10:375.

36. Hansen, E., and A. Harris. 1980. An improved technique for spatial sampling of solutes in shallow groundwater systems. Water Resources Res. 16:827-829.

37. Merritt, W., and P. Parsons. 1960. Sampling devices for water and soil. Disposal of Radioactive Wastes Conf. Proc., Ind. Atomic Energy Agency, Monaco, Nov. 16-21, 1959, 2:329-338.

38. Grisak, G., B. Rehm, and C. Davidson. 1977. Field dispersivity measurements in a shallow granular aquifer. Geologic Assn. of Canada, Annual Management Program with Abstracts 2:22.

39. Pickens J., J. Cherry, G. Grisak, W. Merritt, and B. Risto. 1978. A multilevel device for groundwater sampling and piezometric monitoring. Ground Water 16:322-327.

40. Pickens, J., J. Cherry, R. Coupland, G. Grisak, W. Merritt, and B. Risto. 1981. A multilevel device for ground water sampling. Ground Water Monitoring Review 1(1):48-51.

41. Sudicky, E., and J. Cherry. 1979. Field observations of tracer dispersion under natural flow conditions in an unconfined sandy aquifer. Proc. 14th Canadian Symp. on Water Pollution Res., Toronto, Ontario, Canada, p. 17.

42. Pickens, J., and G. Grisak. 1979. Reply to discussion by Vonhoff, Weyer, and Whitaker, of "A multilevel device for ground-water sampling and piezometric monitoring." Ground Water 17:393-397.

43. Gillham, R., and P. Johnson. 1981. A positive-displacement groundwater sampling device. Proc., First National Ground Water Quality Monitoring Symp. and Exposition, Columbus, OH, p. 40-42.

44. Cherry, J., R. Gillham, G. Anderson, and P. Johnson. 1980. CFB Borden landfill study, Vol. 2: Groundwater monitoring devices. Final Rept. Submitted to the Canadian Dept. of Supply and Services, OSU 78-00195, 34 p.

45. Fenn, D., E. Cocozza, J. Isbister, O. Briads, B. Yare, and P. Roux. 1977. Procedures manual for ground water monitoring at solid waste disposal facilities. U.S. Environmental Protection Agency, EPA/530/SW-611, 269 p.

Gas Lift Samplers

46. Lofy, R., H. Phung, R. Stearns, and J. Walsh. 1977. Investigation of groundwater contamination from subsurface sewage sludge disposal, Vol. I: Project description and findings. U.S. Environmental Protection Agency, 68-01-4166, p. 301-304.

47. Morrison, R., and D. Ross. 1978. Monitoring for ground-water contamination from a hazardous waste disposal site. Proc., 1978 Conf. and Exhibition on Control of Hazardous Material Spills, Miami Beach, FL, Apr. 11-13, p. 281-286.

48. Morrison, R., and P. Brewer. 1981. Air-lift samplers for zone-of-saturation monitoring. Ground Water Monitoring Review 1(1):52-54.

49. Morrison, R., and R. Timmons. 1981. Groundwater monitoring, Part II. Groundwater Digest 4: 21-24.

50. Eleutherius, L. 1980. A rapid in situ method of extracting water from tidal marsh soils. Soil Sci. Soc. Amer. J. 44:884-886.

Pore Pressure Piezometers

51. Russel, H. 1981. Instrumentation and monitoring of excavations. Bull. Assn. Engineering Geologists 18:91-99.

52. Massarsch, K., B. Broms, and O. Sundquist. 1975. Pore pressure determination with multiple piezometer. Proc., In Situ Measurement of Soil Properties, North Carolina State Univ., Raleigh, NC, 1:260-265.

53. Petur Instrument Co., Inc. 1981. Petur model PNE-100 electro-pneumatic piezometer. Petur Instrument Co., Inc., Seattle, WA.

54. Wolff, R., and H. Olsen. 1968. Piezometer for monitoring rapidly changing pore pressures in saturated clays. Water Resources Res. 4:839-843.

55. Petur Instrument Co., Inc. 1981. Petur model P-103, 1:1 pneumatic piezometer. Petur Instrument Co., Inc., Seattle, WA.

56. Wissa, A., R. Martin, and J. Garlanger. 1975. The piezometer probe. Proc., In Situ Measurement of Soil Properties, North Carolina State Univ., Raleigh, NC, 1:536-545.

Well Piezometers

57. Reeve, R. 1965. Hydraulic head. C. Black, D. Evans, J. White, L. Ensimger, and F. Clark (Eds.), Methods of Soil Analyses, Part I. Agronomy, 9 in series monographs. Amer. Soc. Agron., Madison, WI, p. 180-196.

58. Garber, M., and F. Koopman. 1968. Techniques of water-resources investigations of the United States Geological Survey. Methods of Measuring Water Levels in Deep Wells, Dept. of the Interior, Book 8, Chapter 41, p. 23.

59. Metritape, Inc. 1981. Application notes: Metritape gauging for the water market. Metritape, Inc., Concord, MA.

60. Hemond, H. 1982. A low-cost multichannel recording piezometer system for wetland research. Water Resources Res. 18:182-186.

61. Solinst Canada, Ltd. 1979. Hydraulic piezometer. Data Sheet 2W, Solinst Canada, Ltd., Burlington, Ontario, Canada.

62. Solinst Canada, Ltd. 1981. Standpipe piezometer. Data Sheet 1W, Solinst Canada, Ltd., Burlington, Ontario.

63. Hvorslev, M. 1951. Time lag and soil permeability in groundwater observations. U.S. Army Corps of Engineers, Waterways Experim. Sta. Bull. 36:1-50.

64. Lissey, A. 1967. The use of reducers to increase the sensitivity of piezometers. J. Hydrology 5:197-205.

PART III
SAMPLING EQUIPMENT

Many of the monitoring approaches discussed in Part II require auxiliary equipment for sample withdrawal. Samplers designed for this purpose include bailers, suction lift pumps, submersible pumps, air lift pumps, bladder pumps, gas displacement pumps, gas piston pumps, and packer pumps. The selection of a sampler should be based upon the water quality analysis to be performed and the type of unit from which the sample is to be withdrawn.

A. BAILERS

Bailers are samplers which are lowered into a well or tube. The vessel fills with water and the unit is retrieved and transferred into a sample container. The four varieties are the single and double check valve, messenger, and syringe bailers.

1. Single Check Valve Bailers

A check valve at the bottom of the sample chamber seals a single check valve bailer when it is withdrawn from a well. A ball and check seat arrangement or a dart valve (Figure 3.1) can be used. PVC and Teflon® are the most common materials used to fabricate bailers, although stainless steel, acrylic, iron, and ABS have also been used. An all Teflon® construction is the ideal material when sensitive water quality analysis is to be performed.

The bailer shown in Figure 3.2 consists of a press fitted ball check valve and all threaded components. The threads located midpoint on the vessel allow additional lengths of blank casing for increased sampling capacity. The threaded and press fitted assembly also facilitates the total disassembly (Figure 3.3) of the unit for cleaning purposes.

In operation, the single check valve bailer is lowered into the well annulus, water enters the chamber through the bottom, and the weight of the water column closes the check valve upon bailer retrieval. The specific density of the ball should be between 1.4 and 2.0 so that the ball almost sits on the check valve seat during chamber filling. Upon bailer withdrawal, the ball will immediately seat without any sample loss through the check valve. A similar technique involves lowering a sealed sample container within a weighted bottle into the well. The stopper is then pulled from the bottle via a line and the entire assembly is retrieved when the container is full.[2]

Figure 3.1 Dart valve.

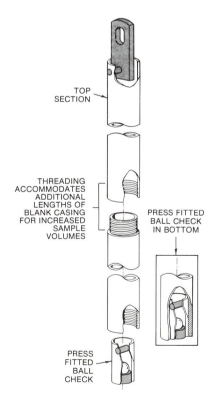

TOP SECTION

THREADING ACCOMMODATES ADDITIONAL LENGTHS OF BLANK CASING FOR INCREASED SAMPLE VOLUMES

PRESS FITTED BALL CHECK IN BOTTOM

PRESS FITTED BALL CHECK

Figure 3.2 Single check valve bailer.[1] (reproduced from TIMCO GEOTECHNICAL PRODUCTS CATALOGUE, 1982, by permission of Timco Mfg., Inc.)

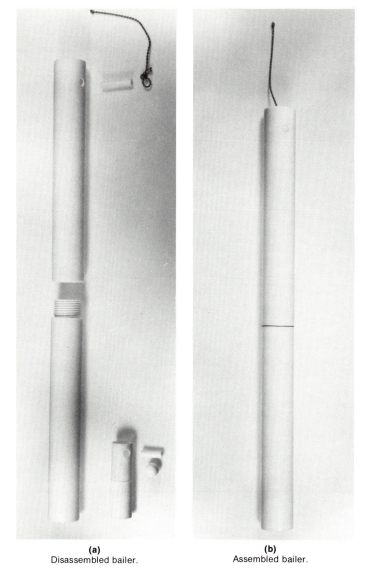

(a)
Disassembled bailer.

(b)
Assembled bailer.

Figure 3.3 Disassembled and complete Teflon® single check valve bailer.
(photographs ©1983, TIMCO MFG., INC., Prairie du Sac, WI)

Single check valve bailers provide a portable and simple means for collecting a water sample. Limitations include the difficulty in ascertaining the point within the water column the sample represents, oxidation of the sample near the surface of the sample, and possible disturbance of the water column by the sampler.

2. Double Check Valve Bailers

The double check valve bailer (also point source bailer) is designed for sampling at a prescribed depth within a water column. In this design, water flows through the sample chamber as the unit is lowered. A Venturi tapered inlet and outlet insures that water passes freely through the unit with minimum disturbance to the water column. When the desired depth is reached, the unit is retrieved. Since the tolerance between each ball and check valve seat is maintained by a pin which blocks the vertical movement of the check ball, both check valves close simultaneously upon retrieval.

The cross section of a double check valve bailer is shown in Figure 3.4. The unit is threaded in the middle for adding blank casing and can be disassembled for cleaning due to the press fitted and threaded construction. The drainage tube in Figure 3.4b is placed in the bottom of the bailer while the other end is attached to an inert length of tubing. The tubing is connected to a two way stoppered bottle into which the nonaerated sample flows.

Double check valve bailers provide a means for collecting a relatively undisturbed water sample within a water column. The availability of an all Teflon® or acrylic (Figure 3.5) construction and the modular fabrication can result in a highly versatile and portable sampler.

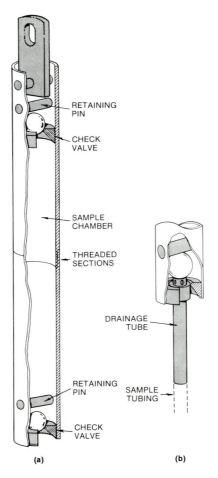

Figure 3.4 (a) Point source bailer and (b) drainage tube. (reproduced from TIMCO GEOTECHNICAL PRODUCTS CATALOGUE, 1982, by permission of Timco Mfg., Inc., Prairie du Sac, WI)

Figure 3.5 Acrylic double check valve bailer. (photograph ©1983, TIMCO MFG., INC., Prairie du Sac, WI)

87

3. Messenger Bailers

A messenger bailer (also refered to as a tube water, thief, or grab sampler) collects point source samples by employing a weight, or similar technique, to close plugs at either end of an open tube. Foerst, Kemmerer, and Bacon samplers are of this variety.[3-5] A drain tube or dart valve can be included in the bottom of a messenger sampler so that a nonaerated sample can be transferred into a collection vessel.

Messenger bailers have limited application in ground water sampling due to the potential introduction of pollutants from the plugs or trip mechanism. Difficulties in handling a messenger and incomplete closing of a plug present additional operational problems.

4. Syringe Bailer

A syringe bailer (sampler) is distinguished from other bailers by the means of water entry.[6] The syringe is lowered into a well and water is drawn into the chamber by activating a plunger via suction. To recover the sample filled syringe, the unit is withdrawn and transferred into a collection bottle or injected directly into an appropriate instrument for water quality analysis.

A syringe sampler is illustrated in Figure 3.6. The plastic or glass syringe can be equipped with a Teflon® plunger from which the handle has been removed. The vacuum and pressure line are used to lower and retrieve the syringe.

The syringe bailer provides an ideal method when used as both a sampler and sample container. Point source sampling is another advantage with this approach, along with its value in collecting samples for volatile organic analysis. The small syringe size is a limitation when large sample volumes are required.

B. SUCTION LIFT PUMPS

Suction lift pumps can be categorized as direct line, centrifugal, and peristaltic. The direct line method involves lowering one end of a plastic tube into a well or piezometer. The surface end of the tube is connected to a two way stoppered bottle, and a manually or auxiliary powered pump is attached to a second tube which leads from the bottle. A check valve is attached between this second line and the vacuum pump to maintain a constant vacuum control.[7-8] This approach has been used to retrieve samples from a tile drain well.

A centrifugal pump approach is similar to the direct line except that a centrifugal pump is connected to the tubing at the surface rather than a vacuum pump. A foot valve is usually attached to the end of the well tubing to assist in priming the extraction tube. The maximum lift is about 4.6 m with this arrangement.[9]

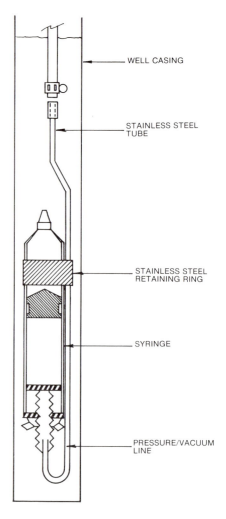

WELL CASING

STAINLESS STEEL TUBE

STAINLESS STEEL RETAINING RING

SYRINGE

PRESSURE/VACUUM LINE

Figure 3.6 Syringe sampler[6] (reproduced from R. Gillham, GROUND WATER MONITORING REVIEW, Volume 2(2), 1982, copyrighted 1982 by Ground Water Monitoring Review)

A peristaltic pump is a self priming, low volume suction pump consisting of a rotor and three ball bearing rollers.[10] Plastic tubing inserted around the pump rotor is squeezed by the rollers as they revolve in a circle around the rotor. One end of the tubing is placed into the well while the other end is connected directly to a two way stoppered flask. As the rotor revolves, water is drawn into the sampling tube and discharged into the collection vessel. A drive shaft connected to the rotor head can be extended so that multiple rotor heads can be attached to a single drive shaft.

These suction lift approaches offer a simple retrieval method for shallow monitoring. All three methods, however, result in sample mixing and oxidation. Degassing also occurs to some extent. A centrifugal pump agitates the sample to a greater degree than the direct line method, although pumping rates of 19 to $151\,\ell$ pm can be attained. A peristaltic pump provides a lower sampling rate ($\approx 3.7\,\ell$ pm) but less agitation than the other two pumps. The withdrawal rate of peristaltic pumps can be carefully regulated by adjusting the rotor head revolution. All three systems can be arranged so that the sample contacts only Teflon® prior to entering the bottle.

C. SUBMERSIBLE PUMPS

A submersible pump consists of a sealed electric motor which powers a helical rotor at high revolutions.[12] Water is transported to the surface by centrifugal action through an access tube.

Submersible pumps provide relatively high discharge rates for water withdrawal at depths beyond suction lift capabilities. A battery operated 3.6 cm diameter unit with a $4.5\,\ell$ pm discharge rate at 33.5 m has been developed (Figure 3.7). Another submersible pump with an outer diameter of 11.4 cm can pump water from 91 m. Pumping rates vary from 26.5 to $53\,\ell$ pm depending upon the depth of the pump.[13]

A submersible pump provides higher extraction rates than most other methods. Considerable sample agitation results, however, in the well and in the turbulent flow of the sample in the tube during transport. The potential introduction of trace metals into the sample from the pump materials also exists. Steam cleaning of the unit followed by rinsing with unchlorinated, deionized water is recommended between sampling when water quality analysis in the parts per million or parts per billion range is required.

D. AIR LIFT PUMPS

An air lift pump collects a water sample by bubbling a gas at depth in the well or tube. Sample transport occurs primarily as a result of the reduced specific gravity of the water being lifted to the surface. Water is forced up a discharge pipe, which may be the outer casing or a smaller diameter pipe inserted into the well.[14-15]

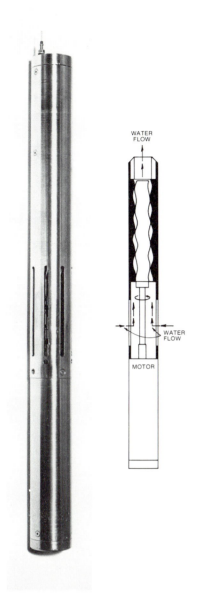

WATER
FLOW

WATER
FLOW

MOTOR

Figure 3.7 Submersible pump.[11] (reproduced from NEW "KECK" SUBMERSIBLE WATER SAMPLING PUMP FOR GROUNDWATER MONITORING, 1981, by permission of Keck Geophysical Instruments, Inc., Okemos, MI)

Similar principles are used in a unit which consists of a small diameter plastic tube perforated at the lower end. The tube is placed within another tube of a slightly larger diameter. Compressed propane is injected into the inner tube. The propane bubbles through the perforations, thereby lifting the water sample via the annulus between the outer and inner tubing.[16] In practice, the eductor line should be submerged to a depth equal to 60% of the total submerged eductor length during pumping.[17] This 60% ratio is considered optimal, although a 30% submergence ratio is adequate.

The source of compressed gas may be a hand pump for depths generally less than 7.6 m. For greater depths, air compressors, pressurized air bottles, and air compressed from an automobile engine have been used. Air lift methods result in considerable sample agitation and mixing in the well, and are not recommended for samples which will be tested for volatile constituents. The considerable pressures required for deep sampling can result in significant redox, pH, and specie transformations. The eductor pipe or weighted plastic tubing is also a potential source of sample contamination.

E. BLADDER PUMPS

Bladder pumps (also referred to as gas squeeze pumps) consist of a flexible membrane enclosed in a rigid stainless steel housing. A strainer or screen attaches below the bladder to filter any material which could clog either of the check valves located above and below the bladder. Water enters the membrane through the lower check valve; compressed gas is injected into the cavity between the housing and bladder. The sample is transported through the upper check valve and into the discharge line. This upper check valve prevents water from reentering the bladder. The process is repeated to cycle the water to the surface.

A variety of designs have been proposed.[18-20] The 4.4 cm pump shown in Figure 3.8 is capable of providing samples (≈ 2.6 to 5.6 ℓ pm) from depths in excess of 76 m. Bladder volumes (e.g., volume per cycle) and sampler geometry can be modified to increase the sampling abilities of the pump. Automated control systems are available to control gas flow rates and pressurization cycles.[21]

Bladder pumps prevent contact between the gas and water sample and can be fabricated entirely of Teflon® and stainless steel. A nearly continuous flow can be attained with the proper cycles. Disadvantages include the large gas volumes required, especially at depth, potential bladder rupture, and the difficulty in disassembling the unit for thorough cleaning.

F. GAS DISPLACEMENT PUMPS

Gas displacement pumps are distinguished from air lift pumps by their method of sample transport. Gas displacement pumps force a column of water under linear flow conditions to the surface without extensive mixing of the pressurized gas and water. A vacuum can also be used with the gas.[22]

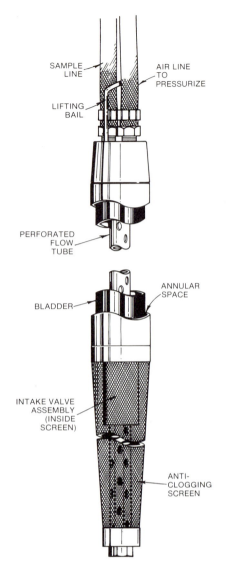

Figure 3.8 Bladder pump.[18] (reproduced from PROCEDURES AND EQUIPMENT FOR GROUND WATER MONITORING, 1981, by permission of Industrial & Environmental Analysts, Inc.)

One gas displacement pump is identical to the in situ sampler described in Section G of Part II. In this application, however, the unit is not buried but is either lowered into the well or installed with the casing which serves as an integral portion of the well.[23-24] In the former, the sampler is lowered to the desired depth and intermittently pressurized. For deeper applications, a transfer vessel (Figure 2.17) is used in conjunction with the pump. A solenoid valve and variable set electronic timer attached to the nitrogen cylinder alternate this cycle. Flow rates of about 2.8 ℓ pm at 36.5 m are possible with a standard 3.7 cm inner diameter by 4.57 cm long pump. Units with a 3.1 cm inner diameter are available although smaller volumes result unless an increased pump length is provided.

For permanent installation within the well, the vessel is threaded to standard PVC casing or screen (Figure 2.15b). Assuming that the pump is threaded to a length of screen above the unit, the well would remain accessible to bailers or other small diameter downhole instruments. A modification to this concept is illustrated in Figure 3.9 where the gas lift pump and associated casing is installed within an existing well thereby forming a rigid sample extraction tube. This retrievable arrangement allows the extraction of greater sample volumes due to the increased inner diameter of the extraction line.

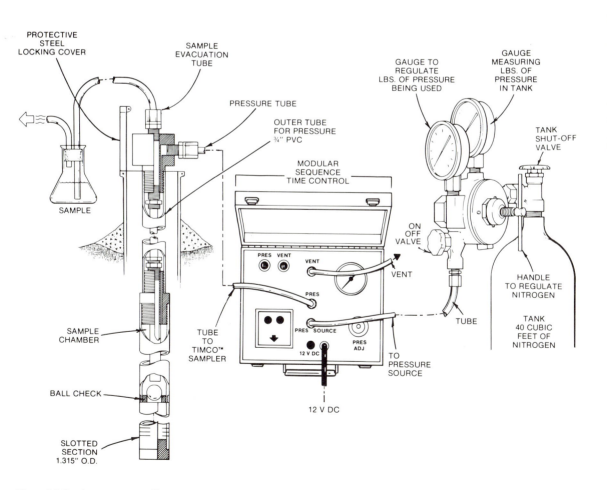

Figure 3.9 Semipermanent gas lift pump arrangement.[25](illustration ©1983, TIMCO MFG., INC.,Prairie du Sac, WI)

A more complicated two stage design constructed of glass with Teflon® check valves has been constructed.[26-27] The unit was designed specifically to test samples for trace level organics. Continuous flow rates up to 2.3 ℓ pm are possible with a 5.1 cm diameter unit.

Gas displacement pumps have been designed with multiple functions. The water sampler in Figure 3.10 provides piezometric measurements with an internally mounted transducer. This unit may also be installed in situ in a manner similar to the gas lift sampler (Section G, Part II). A sampler with its transducer exposed externally for piezometric measurements is illustrated in Figure 3.11. The sensor can activate the gas source at the surface to cause sample chamber pressurization at a predetermined depth. Another design can be used as a water sampler or as a tool for injecting brine or other tracers into a well.[30]

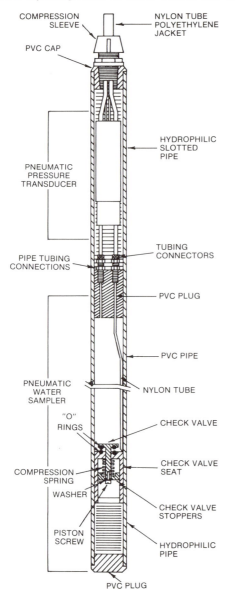

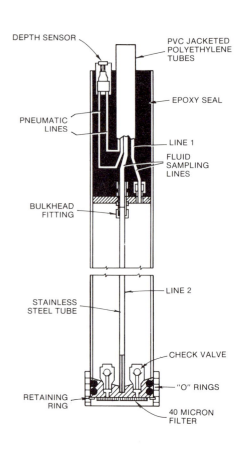

Figure 3.11 Pneumatic water sampler with externally exposed transducer.[29] (reproduced from PETUR LIQUID SAMPLER, 1982, by permission of the Petur Instrument Co. Inc.)

Figure 3.10 Pneumatic water sampler with internal transducer.[28] (reproduced from PNEUMATIC WATER SAMPLER, 1981, by permission of the Slope Indicator Co.)

Gas displacement pumps provide a reliable means for obtaining a highly representative ground water sample. Units constructed of glass or Teflon® with detachable components for easy cleaning should be used when performing sensitive water quality analysis. Correct chamber pressurization minimizes the gas water interface, the degree of mixing, and sample degassing during transport. Most pumps are portable and require only a small gas cylinder for field use.

G. GAS PISTON PUMPS

A double piston pump utilizes compressed air to force a piston to raise the sample to the surface. In the design in Figure 3.12, the stainless steel chamber is between two pistons. The alternating chamber pressurization activates the piston which allows water entry during the suction stroke of the piston and forces the sample to the surface during the pressure stroke. Pumping rates of 0.5 ℓ pm have been reported from 30.5 m; sampling depths of 152 m are possible.[32]

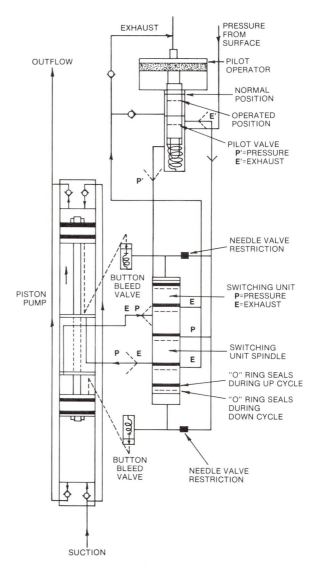

Figure 3.12 Gas driven piston pump.[31]

The gas piston pump provides continuous sample withdrawal at depths greater than is possible with most other approaches. Nevertheless, contribution of trace elements from the stainless steel and brass is a potential problem. Pumping rates at depths less than 152 m are generally slower than with other pumps.

The syringe sampler illustrated in Figure 3.6 can be modified with an intake and discharge line similar to the Figure 2.13 arrangement. Pressure and possibly vacuum are used to activate the plunger in a piston type motion.

H. PACKER PUMPS

A packer pump assemblage provides a means by which two expandable "packers" isolate a sampling unit between them. Since the hydraulic or pneumatic activated packers are wedged against the casing wall or screen, the sampling unit collects water samples only from the isolated portion of the well. The packers deflate for vertical movement within the well and inflate when the desired depth is attained (Figure 3.13).[33]

Figure 3.13 Deflated and expanded packer.[33] (reproduced from TIGRE TIERRA HX PNEUMATIC PACKER, 1981, by permission of Tigre Tierra, Inc.)

Packers are usually constructed from some type of rubber or rubber compound and can be used with submersible, gas lift, and suction pumps.[34-37] A packer pump unit with a vacuum sampler is shown in Figure 3.14.

A packer assembly allows the isolation of sampling points within a well. A number of different samplers can be situated between the packers depending upon the analytical specifications for sample testing. Vertical movement of water outside the well casing during sampling is possible with packer pumps but depends upon the pumping rate and subsequent disturbance. Deterioration of the expandable material will occur with time thereby increasing the possibility of undesirable organic contaminants entering the water sample.

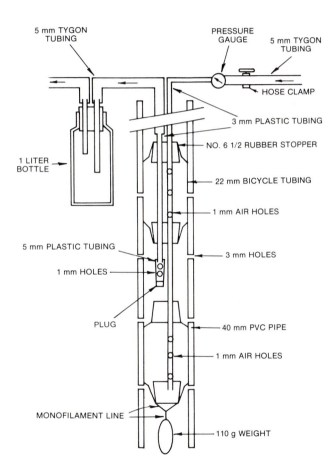

Figure 3.14 Packers with vacuum pump.[36] (reproduced from SOIL SCIENCE SOCIETY OF AMERICA JOURNAL, Volume 44, 1980, page 1120, by permission of the Soil Science Society of America)

REFERENCES — PART III

Bailers

1. Timco Mfg., Inc. 1982. Variable capacity bailer. Timco Geotechnical Products Catalogue, Timco Mfg., Inc., Prairie du Sac, WI.

2. deVera, E., B. Simmons, R. Stephens, and D. Storm. 1980. Samplers and sampling procedures for hazardous waste streams. U.S. Environmental Protection Agency, EPA-600/2-80-018, 51 p.

3. Eijelkamp B. V. 1979. Equipment for soil research. Eijelkamp B.V., Giesbeek, Netherlands, p. 82-83.

4. Huibregtse, K., and J. Moser. 1976. Handbook for sampling and sample preservation of water and wastewater. U.S. Environmental Protection Agency, EPA-600/4-76-049, 255 p.

5. Wood, W. 1975. Techniques of water resources investigations of the United States Geological Survey. Guidelines for Collection and Field Analysis of Ground-Water Samples for Selected Unstable Constituents, Chapter D2, p. 24.

6. Gillham, R. 1982. Syringe devices for ground water sampling. Ground Water Monitoring Review, 2 (2):36-39.

Suction Lift Pumps

7. Allison, L. 1971. A simple device for sampling ground waters in auger holes. Soil Sci. Soc. Amer. Proc. 35:844-845.

8. Willardson, L., B. Meek, and M. Huber. 1972. A flow path ground water sampler. Soil Sci. Soc. Amer. Proc. 36:965-966.

9. Wilson, L. 1980. Monitoring in the vadose zone: A review of technical elements and methods. U.S. Environmental Protection Agency, EPA-600/17-80-134, 180 p.

10. Masterflex. 1981. Masterflex pump catalogue. Barnant Corp., Barrington, IL.

Submersible Pumps

11. W.G. Keck & Associates. 1981. New "Keck" submersible water sampling pump for ground-water monitoring. East Lansing, MI, (now Keck Geophysical Instruments, Inc., Okemos, MI).

12. Koopman, F. 1979. Downhole pumps for water sampling in small diameter wells. U.S. Geological Survey, Open File Rept. 79-1264.

13. McMillon, L., and J. Keeley. 1968. Sampling equipment for ground-water investigations. Ground Water 6:9-11.

Air Lift Pumps

14. Trescott, P., and G. Pinder. 1970. Air pump for small-diameter piezometers. Ground Water 8:10-15.

15. Sommerfeldt, T., and D. Campbell. 1975. A pneumatic system to pump water from piezometers. Ground Water 13:293.

16. Smith, A. 1976. Water sampling made easier with new device. The Johnson Drillers J. July-Aug., p. 1-2.

17. E.E. Johnson, Inc. 1980. Ground Water and Wells. E.E. Johnson, Inc., St. Paul, MN, 440 p.

Bladder Pumps

18. Industrial & Environmental Analysts, Inc. 1981. Procedures and equipment for groundwater monitoring. Industrial & Environmental Analysts, Inc., Essex Junction, VT.

19. Unwin, J. 1982. A guide to groundwater sampling. National Council of the Paper Industry for Air and Stream Improvement. Tech. Bull. 362, p. 67.

20. Leonard Mold and Die Works. 1982. Air squeeze pump. Leonard Mold and Die Works, Denver, CO.

21. Markland Specialty Engineering, Ltd., Bull. 105/78, Automatic sampler controller: Markland model 105 and 2105. Markland Specialty Engineering, Ltd. Etobicoke, Ontario, Canada.

Gas Displacement Pumps

22. Robin, M., D. Dytynyhyn, and S. Sweeney. 1982. Two gas-drive sampling devices. Ground Water Monitoring Review 2(1):63-66.

23. Bianchi, W., C. Johnson, and E. Haskell. 1962. A positive action pump for sampling small bore holes. Soil Sci. Soc. Amer. Proc. 26:86-87.

24. Timmons, R. 1981. Discussion of "An all Teflon bailer and an air-driven pump for evacuating small-diameter ground-water wells" by D. Buss and K. Bandt. Ground Water 19:666-667.

25. Timco Mfg., Inc. 1982. Gas lift Teflon® pump. Timco Geotechnical Products Catalogue, Timco Mfg., Inc.,Prairie du Sac, WI.

26. Tomson, M., S. Hutchins, J. King, and C. Ward. 1980. A nitrogen powered continuous delivery, all-glass-Teflon pumping system for ground-water sampling from below 10 meters. Ground Water 18:444-446.

27. Tomson, N., J. Dauchy, S. Hutchins, C. Curran, C. Cook, and C. Ward. 1981. Groundwater contamination by trace level organics from a rapid infiltration site. Water Res. 15:1109-1115.

28. Slope Indicator Co. 1982. Pneumatic water sampler. Slope Indicator Co., Seattle, WA.

29. Petur Instrument Co., Inc. 1982. Petur liquid sampler. Petur Instrument Co., Inc., Seattle, WA.

30. Idler, G. 1980. Modification of an electronic downhole water sampler. Ground Water 18:532-535.

Gas Piston Pumps

31. Signor, D. 1978. Gas-driven pump for ground-water samples. U.S. Geological Survey, Water Resources Investigation 78-72. Open File Rept. 25. p.

32. Robert Bennett Co. 1982. Bennett sample pumps, model 180-500. Robert Bennett Co., Amarillo, TX.

Packer Pumps

33. Tigre Tierra, Inc. 1981. Tigre Tierra HX pneumatic packer. Tigre Tierra, Inc., Puyullup, WA.

34. Cherry, J. 1965. A portable sampler for collecting water samples from specific zones in uncased or screened wells. U.S. Geological Survey. Prof. Paper 25-C, p. 214-216.

35. Grisak, G., W. Merritt, and D. Williams. 1977. A fluoride borehole dilution apparatus for ground-water velocity measurements. Can. Geotech. J. 14:554-561.

36. Galgowski, C., and W. Wright. 1980. A variable-depth ground-water sampler. Soil Sci. Soc. Amer. J. 44:1120-1121.

37. Cherry, J., and P. Johnson. 1982. A multilevel device for monitoring in fractured rock. Ground Water Monitoring Review 2(3):41-44.

APPENDIX
Metric-English Unit Conversion Table

LENGTH

1 meter (m) = 39.37 inches = 3.28 feet = 1.09 yards
1 kilometer (km) = 0.62 miles
1 millimeter (mm) = 0.03937 inches
1 centimeter (cm) = 0.3937 inches
1 micrometer (μm) = 3.937×10^{-5} = $10^4 \overset{\circ}{\text{A}}$

AREA

1 square meter (m²) = 10.764 feet² = 1.196 yards²
1 square kilometer (km²) = 0.386 miles² = 247 acres
1 square centimeter (cm²) = 0.155 inches²
1 square millimeter (mm²) = 0.00155 inches²

VOLUME

1 kilogram (kg) = 2.205 pounds
1 gram (g) = 0.035 ounces = 15.43 grains
1 milligram (mg) = 0.01543 grains
1 liter (l) = 0.264 U.S. gallons = 1.0566 U.S. quarts

FLOW RATE

1 cubic meter per second (m³/s) = 15,850 U.S. gallons per
    minute = 2,119 cubic feet per minute
1 liter per second (1/l) = 15.85 gallons per minute
1 cubic meter per day (m³/d) = 0.183 gallons per minute

PRESSURE

1 bar = 0.9869 atmospheres = 29.530 inches of Hg at (32°F) =
    14.5 psi
1 centimeter of mercury = 0.01315 atmospheres = 0.3937 inches
    of Hg at (0°C) = 0.1933 psi

TEMPERATURE

degrees Celsium (C) = (5F)/9-17.77
degrees K = degrees C + 273.16

WORK, ENERGY, QUANTITY OF HEAT

1 joule (J) = 2.778×10^{-7} kilowatt hours = 3.725×10^{-7} horsepower
    hours = 0.73756 foot pounds = 9.48×10^{-4} British
    thermal units
1 kilojoule (KJ) = 2.778×10^{-4} kilowatt hours

GLOSSARY

BUFFERING CAPACITY (of a solution) is defined quantitively by the buffering index β which is defined as the slope of a titration curve of pH versus moles of strong bases added ($C\beta$) or moles of strong acid added (C_A):

$$\beta = \frac{dC_B}{dpH} = \frac{-dC_A}{dpH}$$

ABSORPTION penetration of substances into the bulk of a solid or liquid such as the dissolving of a gas by a liquid.

ADSORPTION process by which atoms or molecules or ions are taken up and retained on the surfaces of solids by chemical or physical binding, e.g., the adsorption of cations by negatively charged minerals.

AIR LIFT PUMP consists of two pipes, with one inside the other, to withdraw water from a well. The lower ends of the pipes are submerged and compressed air is delivered through the inner pipe to form a mixture of air and water. This mixture rises in the outer pipe to the surface because the specific gravity of this mixture is less than that of the water column.

ALUMEL nickel based alloy containing 2.5% magnesium, 2% aluminum, 1% silicone and is used chiefly in tyrometric thermocouples.

ALUNDUM (trademark, Norton Company) a porous alumina oxide resembling corundum in hardness. It is manufactured by fusing alumina in an electric furnace and is used chiefly as an abrasive and as a refractive.

ANISOTROPIC characteristic of a material with differing or unequal physical or chemical properties when measured at any given point along its different axes.

BAILER generally an open ended tubular hollow device with a closable valve at or near the bottom end; designed to retrieve water samples from a well or tube.

BAR a unit of pressure equal to one million dynes per square centimeter or 0.98 atmospheres:
(1 bar $\times$ 10^{-1}=MP$_2$).

BENTONITE considered a mineral. An absorptive clay derived from decomposed volcanic ash. It is often used as a grout material (sealant) and is able to absorb large quantities of water and can expand to several times its normal volume; composed primarily of montmorillonite and beidellite.

BISMUTH brittle, grayish white, red-tinged metallic element used in the manufacture of fusible alloys.

BISMUTHINITE mineral bismuth sulfide (Bi_2S_3) occuring in lead-gray masses; an ore of bismuth.

BOURDON GAGE (BOURDON-TUBE-GAGE) consists of a semicircular or coiled, flexible metal tube attached to a gage that records the degree to which the tube is straightened by the pressure of the gas or liquid inside.

BUFFER any substance or mixture of compounds which, when added to a solution, is capable of neutralizing both acids and bases without appreciably changing the original acidity or alkalinity of the solution.

The buffer index indicates the number of moles of acid or base required to produce a prescribed pH change.

BULK DENSITY, SOIL the mass of dry soil per unit bulk volume which is determined before drying to a constant weight at 105°C.

BULK SPECIFIC GRAVITY the ratio of the bulk density of a soil to the mass unit volume of water.

BULK VOLUME the volume, including the soils and the pores, of an arbitrary soil mass.

CAISSON consists of a wood, steel, plastic, or concrete watertight chamber, open at the bottom and containing air under enough pressure to prohibit the entrance of water.

CAPACITANCE electrical property of a non-conductor that allows the storage of energy as a result of electric displacement when opposite surfaces of the non-conductor are maintained at a difference in potential.

CAPACITOR usually consists of two sheets of equally charged conducting metal of opposite signs and separated by a thin film of dielectric material; also called a condensor. A capacitor is used to measure capacitance as defined by:

$$C = \frac{Q}{e}$$

where C = capacitance,
Q = charge in coulombs, and
e = voltage across the capacitor.

CAPILLARY POTENTIAL (MATRIC POTENTIAL) amount of work that must be done per unit quantity of pure water in order to transport reversibly and isothermally an infinitesimal quantity of water (identical in composition to the soil water) from a pool at the elevation and at the external gas pressure of the point under consideration, to the soil water.

CAPILLARY PRESSURE difference in pressure across the interface between two immiscible fluid phases concurrently occupying the interstices of any porous material. This difference is due to the tension of the interfacial surface, the value of which depends upon the surface curvature.

CASTONE dental powder.

CELLULOSE ACETATE acetic acid ester of cellulose.

CENTRIFUGAL PUMP moves a liquid by accelerating it radially outward in an impeller to a surrounding spiral shaped casing.

CERAMIC substance which is composed primarily of clay and similar non-metallic materials, and is formed by firing at high temperatures.

CHROMEL (trademark) one of several nickel based alloys which are resistant to oxidation and to loss of strength due to heat and have a high electrical resistivity and low temperature resistance coefficient. Type P consisting of a 90% nickel and 10% chromium mixture is used for most thermocouples.

COLLIMATOR used to develop a beam of limited cross-section of molecules, atoms or nuclear particles in which the paths of the particles are parallel.

COMPTON EFFECT (COMPTON-DEBYE EFFECT) phenomena which occurs in the x-ray or gamma ray region when a photon interacts with an electron and transfers a portion of its energy to the electron. The electron is ejected from the atom and a new photon of lower energy proceeds from the collision in an altered direction.

CONDUCTANCE power to conduct alternating current; the reciprocal of resistance as expressed in reciprocal ohms or mhos.

CONDUCTION transfer of heat between two parts of a stationary system caused by a temperature difference between the parts.

CONSTANTAN alloy of 55% copper and 45% nickel used for electrical resistance heating and thermocouples.

CONSTANTAN THERMOCOUPLE dual wire junction type temperature detector in which one wire is a constantan alloy and the other either an iron, copper, or Chromel-P alloy.

COULOMB meter-kilogram-second unit of quantity of electricity equal to the quantity of charge transferred in one second across a conductor in which there is a constant current of one ampere.

DARCY'S LAW formulated in 1856 by Henry Darcy of Paris, from extensive work on the flow of water through sand filter beds. (i) law describing the rate of flow of water through porous media. The law is expressed as:

$$Q = kS\left(\frac{H + e}{e}\right)$$

where
Q = volume of water passed in unit time
S = area of the bed
e = thickness of the bed
H = height of the water on top of the bed
"k = coefficient depending on the nature of the sand" and for cases where the pressure "under the filter is equal to the weight of the atmosphere."

(ii) Generalization for three dimensions: The rate of viscous flow of water in isotropic porous media is proportional to, and in the direction of, the hydraulic gradient. (iii) Generalization for other fluids: The rate of viscous flow of homogeneous fluids through isotropic porous media is proportional to, and in the direction of, the driving force.

DEAIRED WATER water from which the air has been removed by vacuum or boiling.

DESORPTION process of removing a sorbed substance by the reverse of adsorption or absorption.

DIELECTRIC CONSTANT (PERMITTIVITY) (ϵ) electrical property as defined by Coulomb's Law:

$$F = \frac{q_1 q_2}{4 \pi \epsilon r^2}$$

where F = force between two charges,
q_1 and q_2 = magnitude of the two charges,
r = distance between the two charges, and
ϵ = permittivity.

ϵ can be expressed as $\epsilon_0 \epsilon_r$ where ϵ_0 is the permittivity of a vacuum in $c^2/N\text{-}m^2$. ϵ_r is referred to as the relative permittivity, or dielectric constant.

DIFFUSION process whereby particles of liquids, gases, or solids intermingle as the result of their spontaneous movement. In dissolved substances, this occurs as the substance moves from a region of higher to one of lower concentration.

DIODE two element (anode and cathode) tube or semiconductor which conducts a current in one direction only.

DISCRIMINATOR circuit that can be adjusted to accept or reject signals of different amplitude or frequency.

DYNAMIC NULL PRINCIPLE realtion used in the accurate measurement of electrical potential; the potential to be measured is balanced by an equal, opposite potential so that no current is drawn from the circuit at which the potential is being measured.

ELASTIC COLLISION the sum of the kinetic energies of translation of the participating collision systems before and after the event are identical.

ELECTROLYSIS change in chemistry where the positively and negatively charged ions migrate to the negative and positive electrodes because an electric current is passed through an electrolyte solution.

ELECTROLYTE chemical compound that will conduct an electric current and will dissociate into ions when dissolved in a suitable solution.

ELECTROMOTIVE FORCE (emf) energy available for conversion from nonelectric to electric form, or vice versa, per unit of charge passing through the source of energy.

FIELD CAPACITY (FIELD MOISTURE CAPACITY) presumed water content remaining in soil 2 or 3 days after a soil has been saturated. This percentage is expressed on a weight or volume basis.

FLUID SCANNING SWITCH consists of a timer circuit and stepper motor which operates on a rotating valve enclosed in a fluid switch wafer. The instrument allows a single pressure transducer to be connected to multiple fluid pressure switches.

FLUX rate of transfer of a quantity (water, heat, etc.) across a surface.

FRITTED GLASS porous glass.

GALVANOMETER instrument used for detecting the existance and/or measuring the strength of a small electric current by movements of a magnetic needle or a coil within a magnetic field.

GAS LIFT mechanical process of lifting a column of water from a well where pressurized gas is used as the lifting agent.

GRAVIMETRIC METHOD procedure for determining the soil moisture content; is expressed as a percentage of dry soil weight.

GROUND WATER that portion of the water below the surface of the ground under conditions where the pressure is greater than atmospheric.

GROUT slurry of cement or bentonite which is used to form an impermeable seal within a cavity or in the pore space of a soil. Other specialized grout materials include epoxy resins, silicone rubbers, lime, fly ash, and bituminous compounds.

HALF-LIFE for radioactive substances, the half-life is the length of time required for half of a given amount of material to disintegrate through radiation. In a chemical sense, the half-life is the time required for one-half of a given material to undergo a chemical reaction.

HYDRAULIC CONDUCTIVITY the proportionality factor in Darcy's Law as applied to the viscous flow of water in soil, i.e., the flux of water per unit gradient of hydraulic potential. If conditions require that the viscosity of the fluid be divorced from the conductivity of the medium, it is convenient to define the permeability of the soil as the conductivity, expressed in:

$$g^{-1} \, cm^3 \, sec.$$

To solve the partial differential equation of the nonsteady-state flow in unsaturated soil, it is convenient to introduce the variable known as the soil water diffusivity.

HYDRAULIC GRADIENT change in static head per unit of distance in a given direction. Unless otherwise specified, the direction is presumed to be that of the maximum rate of decrease in head. The gradient of the head is a mathematic expression which refers to the vector (Δh or grad h). The vector's magnitude is equal to the maximum direction in which the maximum rate of increase occurs. The gradient of the head and hydraulic gradient are equal but of opposite sign.

HYGROMETRIC (DEW POINT) METHOD technique of measuring water potential by determining the dew point depression temperature.

HYSTERESIS a change in the shape of a soil water characteristic curve which differs depending upon whether soil sorption or desorption occurs.

IDEAL GAS CONSTANT (R) proportionality constant with a numerical value depending on the units in which pressure and volume are measured. If pressure is expressed in atmospheres and volumes in liters, then R = 0.082054 liter atm deg^{-1} mole^{-1}.

INFILTRATION in soil science, the downward entry of water into the soil.

INFILTRATION RATE (INFILTRATION CAPACITY) rate at which water enters the soil. It has the dimensions of velocity (i.e.,$m^3m^2sec^1$=m sec^1).

INTERSTITIAL WATER (PORE WATER) water which occupies an open space between solid soil particles.

INVAR (trademark) alloy with a very low coefficient of expansion at atmospheric temperatures; consists of 36% nickel, 0.35% manganese, and the remainder an iron and carbon mixture.

ISOTROPIC SOIL exhibits identical properties in all directions from a given point.

LYSIMETER used to measure percolation and leaching losses from a column of soil under controlled conditions, or for measuring gains and losses by collecting soil pore water via suction in the unsaturated zone. Lysimeters are capable of retaining the accumulated water within the sampling vessel.

MANOMETER used to measure fluid or vapor pressures; consists of a tube filled with a liquid so that the level of the liquid is determined by the fluid pressure, and the height of the liquid may be read from a scale. Double leg liquid column gages, pressure transducers, Bourdon gages, and strain gages have also been used to measure fluid pressure.

MATRIC POTENTIAL (CAPILLARY POTENTIAL) amount of work that must be done per unit quantity of pure water in order to transport reversibly and isothermally an infinitesimal quantity of water (identical in composition to the soil water) from a pool at the elevation and at the external gas pressure of the point under consideration, to the soil water.

MATRIX solid portion of a porous material.

MONEL (trademark) corrosion resistant alloy which contains 67% nickel and 30%copper and the remaining 3% is either aluminum or silicone.

MONTMORILLONITE aluminosilicate clay mineral with 2:1 expansible layer structure (two silicone tetrahedral sheets enclosing an aluminum octahedral sheet). Considerable expansion may be caused along the C-axis by water moving between the silica layers of contiguous units.

OHM resistance through which a difference in one volt will produce a current of one ampere.

OSCILLATOR instrument used to produce an alternating current by converting a direct current source to a periodically varying electrical output.

OSMOTIC POTENTIAL amount of work that must be done per unit quantity of pure water in order to transport reversibly and isothermally an infinitesimal quantity of water from a pool of pure water, at a specified elevation and at atmospheric pressure, to a pool of water identical in composition to the equilibrium soil solution (at the point under consideration), but in all other respects being identical to the reference pool.

OSMOTIC PRESSURE pressure to which a pool of water, identical to the soil water, must be subjected through a semipermeable membrane in order to be in equilibrium with a pool of pure water (semipermeable means permeable only to water).

PELTIER COEFFICIENT ratio of the theoretical rate at which heat is evolved or absorbed at a junction of two dissimilar metals (Peltier effect) to the current passing through the junction.

PELTIER EFFECT heat which is evolved or absorbed after allowance for resistance at the junction of two dissimilar metals (as in a thermocouple) carrying a small current and which is dependent upon the direction of the current.

PERCOLATION downward movement of water through soil; especially the downward flow to water in saturated or nearly saturated soil at hydraulic gradients of the order of 1.0 or less.

PERISTALTIC PUMP a low volume suction pump. The compression of a flexible tube by a rotor results in the suction development.

PERMEABILITY capacity of a material to be penetrated, saturated, or diffused through by a fluid.

PERMEABILITY, SOIL (i) ease with which gases, liquids or plant roots penetrate or pass through a bulk mass of soil or a layer of soil. Since different soil horizons vary in permeability the particular horizon under question should be designated. (ii) The property of a porous medium itself that relates to the ease with which gases, liquids, or other substances can pass through it; the "K" in intrinsic permeability.

pF (obsolete) common logarithm of the height of a water column in centimeters equivalent to the soil moisture tension. pF is often expressed as suction (negative pressure) or in a potential energy (per unit mass) basis.

pH used to express the intensity of acidic or alkaline condition of a solution; the symbol for the logarithm of the reciprocal of hydrogen ion concentration in gram atoms per liter.

PIEZOMETER any of several instruments used to measure the liquid pressure in soil or other porous material.

PISTON PUMP consists of a piston rod, cylinder, and check valve. Piston pumps force water to the surface through positive displacement.

PLEXIGLASS thermoplastic polymer of methyl methacrylate that is light weight, weather resistant, and has thermoplastic memory.

PORE WATER (INTERSTITIAL WATER) water which occupies an open space between solid soil particles.

POROSITY ratio (usually expressed as a percentage) of the volume of voids of a rock, soil or other porous material to the total volume of the rock or soil mass.

POTENTIOMETER instrument used for the measurement of an electrical potential by comparison with a known potential difference.

PSYCHROMETER instrument used to measure soil humidity by means of a thermocouple which is cooled below the dew point by means of the Peltier effect. The water on the thermocouple evaporates causing the junction temperature to be depressed below the ambient temperature. The wet bulb temperature depression persists until all the water has evaporated; the thermocouple then returns to the ambient temperature.

PULSE HEIGHT ANALYZER instrument used to measure the pulse height whenever the amplitude of the pulse is proportional to the energy dissipation in a detector.

PYCNOMETER a container use to determine the density of a liquid or solid having a specific volume. A thermometer is often included to indicate the temperature of the contained substance.

REDOX abbreviation for an oxidation and reduction reaction in which oxidation is defined as an increase in oxidation number and reduction as a decrease in oxidation number.

SCANNING CURVES curves between the wetting and drying curves in a soil moisture characteristic curve which describes the wetting and drying values between different intermediate wetness values.

SCINTILLATION COUNTER (SCINTILLATION DETECTOR) instrument used to detect ionizing radiation; consists of combined scintillator and photomultiplier.

SCINTILLATOR instrument used to convert radioactive energy into light. When an ionizing particle is absorbed in any one of several transparent scintillators, some of the energy acquired by the scintillator is emitted as a pulse of visible or near visible ultraviolet light.

SEEBECK EFFECT phenomen which occurs when two dissimilar metals are welded at one end; heating of this welded juncture results in a voltage on the free ends.

SLURRY free flowing, pumpable suspension of fine solid material in a liquid.

SODIUM ADSORPTION RATIO (SAR) relation between soluble sodium and soluble divalent cations which can be used to predict the exchangeable sodium percentage of soil equilibrated with a given solution. SAR is defined by:

$$SAR = [Na^+] / \sqrt{\frac{[Ca^{+2}] + [Mg^{+2}]}{2}}$$

SOIL BULK DENSITY (DRY) ratio of the mass of dried soil to its total volume.

SOIL BULK DENSITY (WET) expression of the total mass of moist soil per unit volume.

SOIL MOISTURE CONTENT amount of water contained in the soil, generally expressed as a percentage.

$$\% \text{ Moisture Content} = \frac{\text{Mass of Water in the Soil}}{\text{Mass of Dry Soil}} \times 100$$

SOIL MOISTURE POTENTIAL (TOTAL POTENTIAL) amount of work that must be done per unit quantity of pure water in order to transport reversibly and isothermally an infinitesimal quantity of water from a pool of pure water, at a specified elevation and at atmospheric pressure, to a pool of water identical in composition to the equilibrium soil solution (at the point under consideration), but in all other respects, being identical to the reference pool.

SOIL MOISTURE TENSION the equivalent negative pressure in the soil water. It is equal to the equivalent pressure that must be applied to the soil water to bring it to hydraulic equilibrium through a porous permeable wall or membrane with a pool of water of the same composition. The pressures used and the corresponding percentages most commonly determined are: fifteen-atmosphere percentage, fifteen bar percentage, one-third-atmosphere percentage, one-third-bar percentage, and sixty-centimeter percentage.

SOIL PORES that part of the bulk volume of soil not occupied by soil particles; interstices; voids.

SOIL SALINITY amount of soluble salts in a soil. The conventional measure of soil salinity is the electrical conductivity of a saturation extract.

SOIL TEXTURE relative proportions of the various soil separates in a soil as described by the classes of soil texture shown below. The textural classes may be modified by the addition of suitable adjectives when coarse fragments are present in substantial amounts; for example, "stony silt loam," or "silt loam, stony phase." The sand, loamy sand, and sandy loam are further subdivided on the basis of the proportions of the various sand separates present.

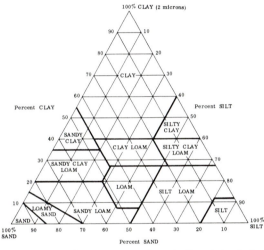

Graph showing the percentages of sand, silt, and clay in the soil classes.

SOIL WATER CHARACTERISTIC CURVE (SOIL MOISTURE RETENTION CURVE) experimentally derived graph showing the soil moisture percentage (by weight or volume) versus applied tension (or pressure). Points on the graph are usually obtained by increasing (or decreasing) the applied tension or pressure over a specified range.

SPECIFIC CONDUCTANCE electrical conductivity of a water sample at 25°C as expressed in μ ohms per centimeter.

SPECIFIC GRAVITY ratio of the mass of a body to the mass of an equal volume of water at 4°C.

SPECIFIC YIELD ratio of the volume of water which a rock or soil, after being saturated, will yield by gravity to the volume of the rock or soil.

SURFACE TENSION free energy in a liquid surface produced by the unbalanced inward pull exerted by the underlying molecules upon the layer of molecules at the surface.

TENSILE STRENGTH resistance of a material to longitudinal stress, measured by the amount of longitudinal stress required to rupture a material or to detach or pull one object from another.

TENSIOMETER a device used to measure the in situ soil water matric potential (or tension); a porous, permeable cup connected to a rigid tube which is attached to a manometer, vacuum gage, pressure transucer, or other pressure measuring device.

THERMAL emf net emf set up in a thermocouple under zero current conditions and which represents the algebraic sum of the Peltier and Thomson emf.

THERMISTOR electronic semiconductor of fused metal oxides located between the conductor and insulator. The electrical resistance varies with temperature.

THERMOCOUPLE union of two dissimilar metallic conductors which create a thermoelectric current. It is used to determine the temperature of a third substance by connecting it to the junction of the metals and measuring the electromotive force produced.

THOMSON emf product of the Thomson coefficient (the rate at which heat is absorbed or evolved reversibly in a thermoelement, per unit temperature difference per unit current) and the temperature difference across a thermoelement.

TORQUE effectiveness of a force to produce rotation about a center, measured by the product of the force and the perpendicular from the line of action of the force to the center about which the rotation occurs. Torque is usually measured at one foot radii.

TORTUOSITY dimensionless geometric parameter of porous media which is the average ratio of the actual erratic path to the apparent or straight flow path; the non-straight nature of soil pores.

TRANSDUCER device that is actuated by power from one system and retransmits it, often in a different form, to a second system.

TREMIE PIPE small diameter pipe with a valve arrangement at the bottom end and a funnel-like top into which grout is poured.

VADOSE ZONE geologic column extending from the ground surface down to the upper surface of the principle water bearing formation.

WATER CONTENT amount of water stored within a porous matrix as expressed on a volume per unit volume or mass per unit mass of solid basis.

WHEATSTONE BRIDGE four arm bridge circuit used for measuring the electrical resistance of an unknown resistor by comparison to the energy dissipation in a detector.

ZONE OF SATURATION zone below the water table in which all interstices are occupied by ground water.

INDEX